Acta Bot. Fennica 156: 1–21 ISBN 951-9469-50-8 ISSN 0001-5369
Helsinki 4 September 1996

Bryophyte flora of the Huon Peninsula, Papua New Guinea. LVII. *Fleischerobryum* and *Philonotis* (Bartramiaceae, Musci)

Timo Koponen & Daniel H. Norris

Koponen, T. & Norris, D. H., Department of Ecology and Systematics, Division of Systematic Biology, P.O. Box 7, FIN-00014 University of Helsinki, Finland

Received 30 August 1995

The family Bartramiaceae has six genera, *Bartramia* Hedw., *Breutelia* (Bruch & Schimp.) Schimp., *Conostomum* Sw., *Fleischerobryum* Loeske, *Leiomela* (Mitt.) Broth. and *Philonotis* Brid. in Western Melanesia. *Fleischerobryum longicolle* (Hampe) Loeske is for the first time recorded for Western Melanesia. The genus *Fleischerobryum* is revised worldwide to have only two species, *F. longicolle* and *F. macrophyllum* Broth. Six species of *Philonotis* are dealt with in this paper. They are *P. calomicra* Broth., *P. hastata* (Duby) Wijk & Marg., *P. mollis* (Dozy & Molk.) Mitt., *P. runcinata* C. Müll. ex Aongstroem, *P. thwaitesii* Mitt. and *P. vescoana* (Besch.) Par. Seventeen names are reduced as synonyms and 10 lectotypes are selected (indexed). *Philonotis calomicra* is for the first time recorded for the Philippines and Samoa, *P. runcinata* for Fiji, Philippines, New Caledonia, Sabah and Western Melanesia, *P. thwaitesii* for Tonga and *P. vescoana* for Fiji, Western Melanesia and Vanuatu. Key to Western Melanesian *Philonotis* is given with the descriptions and illustrations of the species. Distributions in Western Melanesia are given along with the total range. The work includes taxonomic and nomenclatural discussions.

Key words: Bartramiaceae, *Fleischerobryum*, nomenclature, Papua New Guinea, *Philonotis*, Solomon Islands, taxonomy, West Irian

CONTENTS

INTRODUCTION

This paper is part of a series dealing with the bryoflora of Western Melanesia (Papua New Guinea, West Irian, and the Solomon Islands). It is based upon the 1981 collection of T. Koponen and D. H. Norris on the Huon Peninsula, Papua New Guinea, and on extensive herbarium study of other Western Melanesian specimens. For the Norris and Koponen collections we list the collection localities in abbreviated form; we have also codified the presentation of the geographic ranges. Explanation of this format is in the first and seventh parts of this series (Koponen & Norris 1983, Norris & Koponen 1985a; see also Norris & Koponen 1990). In addition to our own specimens, we cite here specimens sent from W. A. Weber (COLO) and H. Streimann (CBG).

Koponen (1990, 1993) and Koponen *et al.* (1992) have dealt with the project and its results. The previous parts of the Huon Peninsula series are listed in Koponen *et al.* (1991) and Koponen (1993b). The previous paper in the series is Pócs *et al.* (1995), see also Pócs *et al.* (1994).

All type and other specimens mentioned here have been seen and verified unless stated otherwise. The nomenclatural references have been checked against the original literature.

FAMILY BARTRAMIACAE SCHWAEGR. 1830

Bartramiaceae is a family of ca. 10 genera. Its limitation and the genera accepted in it have been rather well established. Brotherus (1924) included 9 genera, *Anacolia* Schimp., *Bartramia* Hedw., *Bartramidula* Bruch & Schimp., *Breutelia* (Bruch & Schimp.) Schimp., *Conostomum* Sw., *Exodokidium* Card., *Leiomela* (Mitt.) Broth., *Philonotis* Brid. and *Plagiopus* Brid. Of them *Exodokidium* was synonymized with *Bartramia* by Newton (1973, 1974). Brotherus (1924) placed *Philonotis longicolle* in the section *Pseudo-philonotis* Fleisch. and omitted the genus *Fleischerobryum* described by Loeske (1910). Later he accepted it (Brotherus 1926). Magill (1987) described the genus *Quathlamba*, and Griffin and Buck (1989) the genus *Flowersia*, a segregate from *Anacolia*. Griffin and Buck (1989) discussed the taxonomy and phylogeny of Bartramiaceae and considered *Bartramidula* to be polyphyletic and synonymized it with *Philonotis*. We agree that the characters used for separating *Philonotis* from *Bartramidula*: the generally small size, upright capsule and the absence of a peristome are not sufficient for generic separation. As to the peristome, there is a continuous variation from species with complete double peristomes through reduction of endostome to eperistomate species.

Griffin and Buck (1989) further included the genus *Catoscopium* Brid. in Bartramiaceae, but Akiyama and Nishimura (1993) doubted the correctness of this placement.

In Western Melanesia the family Bartramiaceae has six genera, *Bartramia*, *Breutelia*, *Conostomum*, *Fleischerobryum*, *Leiomela* and *Philonotis*. In this paper we deal only with *Fleischerobryum* and *Philonotis*.

Key to the species of *Fleischerobryum* and *Philonotis* in Western Melanesia

The key works best with the stem leaves.

1. Leaves from broad ovate base abruptly tapering into arcuate apex; basal cells wide and thin-walled, bulging in proximal end; juxtacostal mid-leaf cells with central mammilla/papilla .... *Fleischerobryum*
— Leaves linear-lanceolate to ovate; basal cells distinctly mammillose or papillose in distal end or rarely in proximal end, or narrow high papillose; cells with central papillae not present 2
2. The majority of mammillae/papillae of basal leaf cells proximal; capsule globose, erect, smooth; peristome and enstome reduced .... 1. *P. calomicra*
— The majority of mammillae/papillae of basal leaf cells distal; capsule short ellipsoid or gibbous, inclined to horizontal, ribbed; peristome and endostome ± complete 3
3. Basal leaf cells or all leaf cells thin-walled, quadrate, rectangular or rhomboidal, nearly smooth or low mammillose; costa not reaching the apex or percurrent, (often excurrent on stem leaves); marginal teeth blunt .... 2. *P. hastata*
— Leaf cells often with thickened walls, linear to elongate with distinct papillae all over the leaf or on distal leaf; costa percurrent or excurrent; marginal teeth sharp . 4
4. Leaf borders plane, or only one leaf border recurved to base 5
— Both leaf borders recurved 6
5. Leaves soft, erect spreading to spreading when dry; basal leaf cells lax, long elongate, distinctly broader than distal leaf cells; basal marginal cells rectangular, smooth; capsule glossy .... 3. *P. mollis*
— Leaves stiff, imbricate when dry; basal leaf cells quadrate to short elongate, nearly as narrow as distal leaf cells; basal marginal cells quadrate, crenulate; capsule dim .... 4. *P. runcinata*

6 Basal leaf cells with distinct papillae, not much broader (– 2 ×) than distal leaf cells, quadrate to short rectangular .... 5. *P. thwaitesii*
— Basal leaf cells lax, with low mammillae, distinctly broader (more than 2 ×) than distal leaf cells, rectangular to rhomboidal .... 6. *P. vescoana*

Genus **Fleischerobryum** Loeske 1910

Fleischerobryum is distinguished from *Philonotis* by its long capsule which has a long neck. Basal leaf cells are larger and all leaf cells less papillose than in most *Philonotis* species. Four species has been accepted in the genus: *Fleischerobryum longicolle* (Hampe) Loeske, *F. macrophyllum* Broth., *F. eurybrochis* (Ren. & Card.) Fleisch., and *F. wallisii* (C. Müll.) Loeske. Two latter species are synonymized in this paper, so that only two species remain. *Fleischerobryum longicolle* is widely distributed in Southeast Asia. *Fleischerobryum macrophyllum* is known from the Philippines, Taiwan (Redfearn *et al.* 1996), and the Himalayas (Gangulee 1974). On the basis of the type specimen (Luzon, Bontoc subprovince, Mt. Polis, 1920 *Ramos & Edano 38239*, H-BR) it differs from *F. longicolle* by linear-lanceolate leaves, longer and narrower basal leaf cells, and by only low mammillose cells; mid-leaf juxtacostal cells do not have central mammillae/papillae in *F. macrophyllum*.

Fleischerobryum longicolle (Hampe) Loeske (Fig. 1)

Morph. Syst. Stud. Laubm. 127. 1910. — *Bartramia longicollis* Hampe in C. Müll., Syn. Musc. Frond. 1: 478. 1848. — *Philonotis longicollis* (Hampe) Mitt., J. Linn. Soc. Bot. Suppl. 1: 64. 1859.

Plants to 5 cm tall, dark green, in loose tufts; stems kept adhering by dense tomentum; macronematal apparatus abaxial between costa and lateral margin of leaf; branched with a few subfloral innovations; stem cross-section with hyalodermis, one layer of substereidal cortical cells and with thin-walled medullar and central cells, distinct central strand; leaves closely appressed when dry, erect spreading when moist, concave, arcuate and secund, ca 0.8 mm wide and 2 mm long, from broad ovate base abruptly tapering to acuminate apex, border recurved, leaf margin entire in recurved part, crenulate by protruding cell corners towards apex; costa brownish, ca 50 µm at base, nearly smooth, percurrent or shortly excurrent; leaf cells in ovate base, 30–40 × 50–100 µm, thin-walled and broadly mammillose at proximal cell end; cells in acuminate apex much smaller, 5–10 × 15–40 µm, rectangular, with firm walls, mammillose/papillose in both ends. On juxtacostal mid-leaf cells central mammilla often present (not observed in New Guinea specimen).

Dioicous. Perichaetia terminal, inner perichaetial leaves hardly differentiated from stem leaves. Sporophytes not present in New Guinea specimen, but the capsule is elongate with a long neck.

Illustrations: Dozy & Molkenboer 1845–48: tab. CXXIX, as *Philonotis longicollis*; Brotherus 1924: 467 (fig. 410, as *P. longicollis*); Gangulee 1974: 1100 (fig. 534); Iwatsuki & Mizutani 1972: pl. 18 (fig. 256); Noguchi 1989: 555 (fig. 246B).

Fleischerobryum longicolle can be separated from the local species of *Philonotis* by its rather large size of leaves, which have broad ovate base and long abruptly tapering apical part. The combination of the leaf cell areolation also is different from all *Philonotis*: Area of wide thin-walled cells fills nearly all the basal ovate half of the leaf, basal cells are low mammillose in the proximal end of the cell; the leaf cells in the narrow apex are narrow and with firm walls, and they are low mammillose in the distal end. In the mid-leaf there are several juxtacostal cells which have central papillae. Leaf cells with central mammilla is a character of some species of *Philonotis* sect. *Philonotis*, but species of that section do not occur in New Guinea. Similar wide and thin-walled basal cells occur in *P. hastata* but it has ovate leaf shape, and in *P. vescoana* which has the cells much more mammillose/papillose than in *Fleischerobryum longicolle*.

Only one specimen of *Fleischerobryum longicolle* was collected during the Koponen-Norris expedition.

Range on the Huon Peninsula. *6k*. On boulder in deep S-sloping valley with scattered scrub and springs, alt. 3 500–3 570 m, *Koponen 32692*.

Range in Western Melanesia (first record): Papua New Guinea. *12*. The present report.

Total range: As 2: Chi Ja. As 3: In. As 4: Ind PNG.

Genus **Philonotis** Brid. 1827

The species of the genus *Philonotis* in Southwest Asia and the West Pacific have been revised and

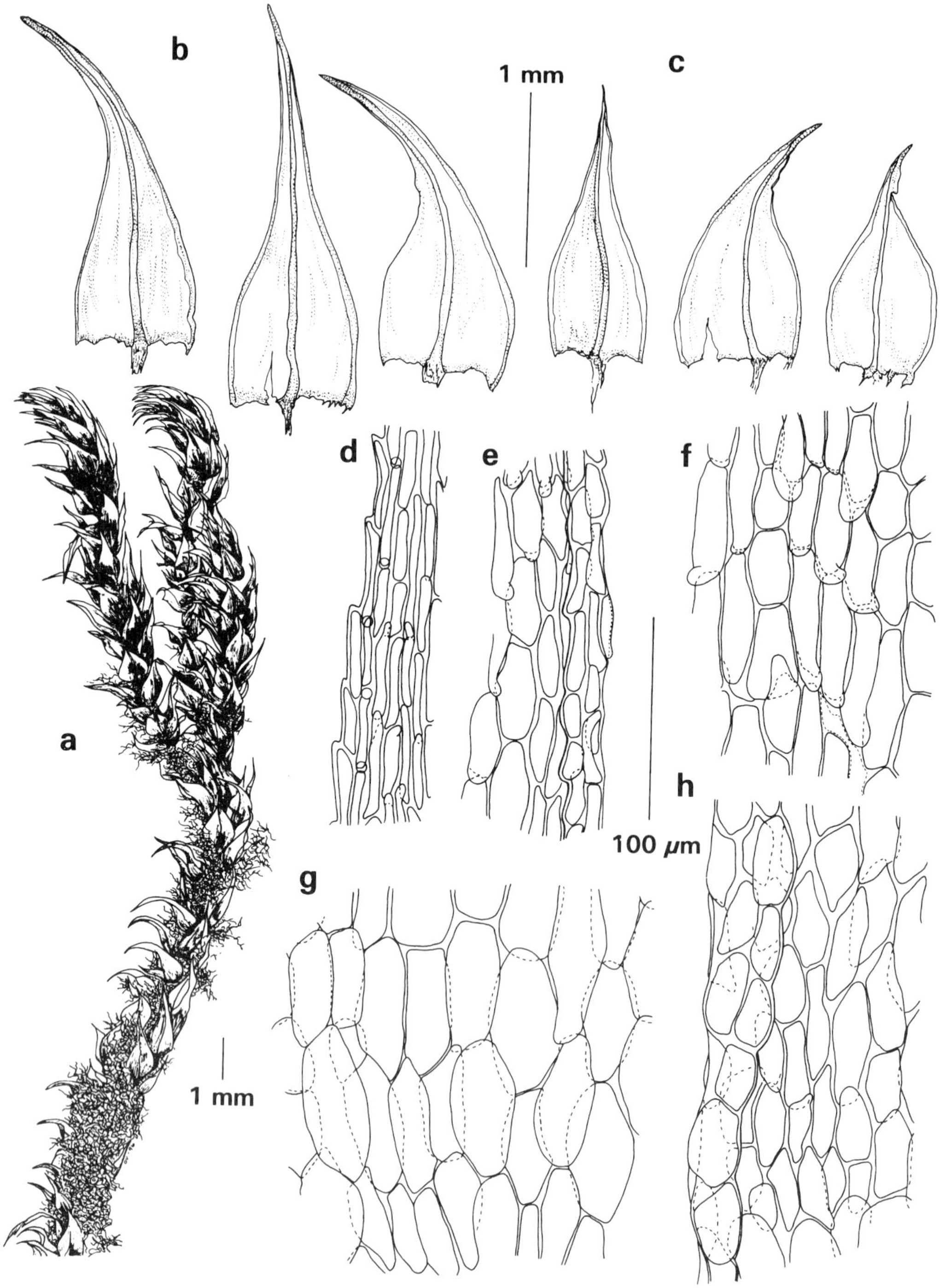

Fig. 1. *Fleischerobryum longicolle* (Hampe) Loeske (from *Koponen 32692*, H). — a: Growth habit. — b: Stem leaves. — c: Branch leaves. — d: Distal leaf border. — e: Mid-leaf border. — f: Mid-leaf cells. — g: Basal leaf cells. — h: Basal leaf border. — Use the 1 mm scale below for a, the 1 mm scale above for b and c, and the 100 μm scale for d–h.

treated taxonomically by several authors (Fleischer 1904, Kabiersch 1938, Bartram 1939, Ochi 1962, 1963, Scott *et al.* 1978, Whittier 1978). Sections and subsections within the genus were used by Fleischer (1904), Brotherus (1924), and Kabiersch (1938) created a subdivision of the genus using the categories section and subsection. This classification has not been followed by subsequent authors. For instance, Ochi (1963) noted that the sections and subsections adopted by Kabiersch (1938) and Sakurai (1941) seem to contradict each other. He also found that the sectional characters were not clear-cut with the result that closely related species are placed in different sections. We found the same, for instance, *Philonotis socia* Mitt. was placed in the section *Sociae* and *P. revoluta* Bosch & Sande Lac. in the section *Revolutae* by Kabiersch (1938), although they are actually synonymous. The basic reason for this confusion was probably due to the fact that early authors described too many species. Ochi (1962, 1963) synonymized numerous taxa as did Whittier (1987). We do agree with most of their revisions, and propose some new synonymies. To establish the nomenclature of New Guinea *Philonotis*, we studied most of the types of *Philonotis* species from neighbouring areas. This led to synonymizing of 18 names. Some of these were reported previously by Koponen (1993a), but here we cite all relevant data, and so we present all these as *syn. nov.*

The generic limits of *Philonotis* are reasonably well established. *Philonotis* sect. *Bartramidula* is the only section of *Philonotis* which has been treated at the generic level by some authors (e.g. Smith 1978, Magill 1987). The other closely related genus, *Fleischerobryum* was accepted by Ochi (1962, 1963) and Griffin and Buck (1989).

1. *Philonotis calomicra* Broth. (Figs. 2, 8)

in Schum. & Lauterb., Fl. Deutsch. Schutzgeb. Südsee 88. 1900. — Type: Papua New Guinea. Madang, Suor Mana an Felsen, 400 m, 9. VI. 1896 *Lauterbach 2304b* (holotype, H-BR).

P. imperfecta Bartr., Philippine J. Sci. 61: 241. 1936, *syn. nov.* — *Bartramidula imperfecta* (Bartr.) Iwats. & Tan, Kalikasan 9: 271. 1980. — Type: Malaysia. Sabah, British North Borneo, Mount Kinabalu, Dallas, damp place, 3 000 ft., 7. VI. 1932 *Clemens s. n.* (holotype, FH-Bartram).

Plants 0.5–1.5 cm tall, pale green, in loose tufts; stems kept adhering by dense tomentum, branched with subfloral innovations, branches arcuate; epidermis in stem cross-section with thin outer cell walls (hyalodermis), one layer of substereidal cortical cells and with thin-walled medullar and central cells, distinct central strand; leaves closely appressed when dry, erect spreading when moist, plane, straight or curved, 0.2–0.4 × 0.8–1.3 mm, acuminate from ovate base, usually broadest above base and slightly narrowed at insertion, apex stiff; border plane or recurved, leaf margin with entire base, geminate towards apex; costa stout, brownish, 25–40 µm at base, papillosely toothed on both sides, percurrent or shortly excurrent, cross-section with thin-walled papillose-mammillose epidermis, conductive strand and dorsal substereidal cells; midleaf cells linear, 7–10 × 30–52 µm, with slightly thickened walls and distinct papilla at upper cell end on ventral surface and with lower mamilla on lower cell end on dorsal surface, apical cells longer and narrower, basal cells wider and shorter, thin-walled, 7–12 × 12–57 µm, often quadrate near border, rectangular near costa; mammillae on proximal cell ends more distinct than distal mammillae.

Dioicous. Perichaetia terminal, inner perichaetial leaves hardly differentiated from stem leaves; perigonia gemmiform, perigonial leaves ca. 1.5 mm long, base of inner leaves cochleariform, apex abrubtly contracted to acute-acuminate apex; seta 1.5–2.5 cm long, capsule erect, smooth when dry, globose to short ellipsoid, ca. 1 mm in diameter, mouth small, operculum flat or with small mammilla; exothecial cells thinwalled, ± isodiametric, 25–50 µm; peristome smooth, deeply inserted; base of peristome teeth brown, trabeculate, complete, or apical part pale, fragile and perforated; endostome fragile and therefore mostly wanting. Spores ca. 25 µm, papillose in light microscope.

Illustrations: Bartram 1939: pl. 12 (fig. 197, as *Philonotis tjibodensis*); Iwatsuki & Noguchi 1975: 325 (fig. 6, as *P. imperfecta*).

When fertile *Philonotis calomicra* is easy to separate from other New Guinea species of *Philonotis* by the upright, smooth and globose capsules. If the capsules are absent, the mammillae on proximal ends of basal leaf cells which are more pronounced than distal papillae, can be used as a diagnostic character. This seems to be a character separating the species of the section *Bartramidula* from most other *Philonotis* species. The endostome is so fragile that in old capsules only the exostome remains. The other spe-

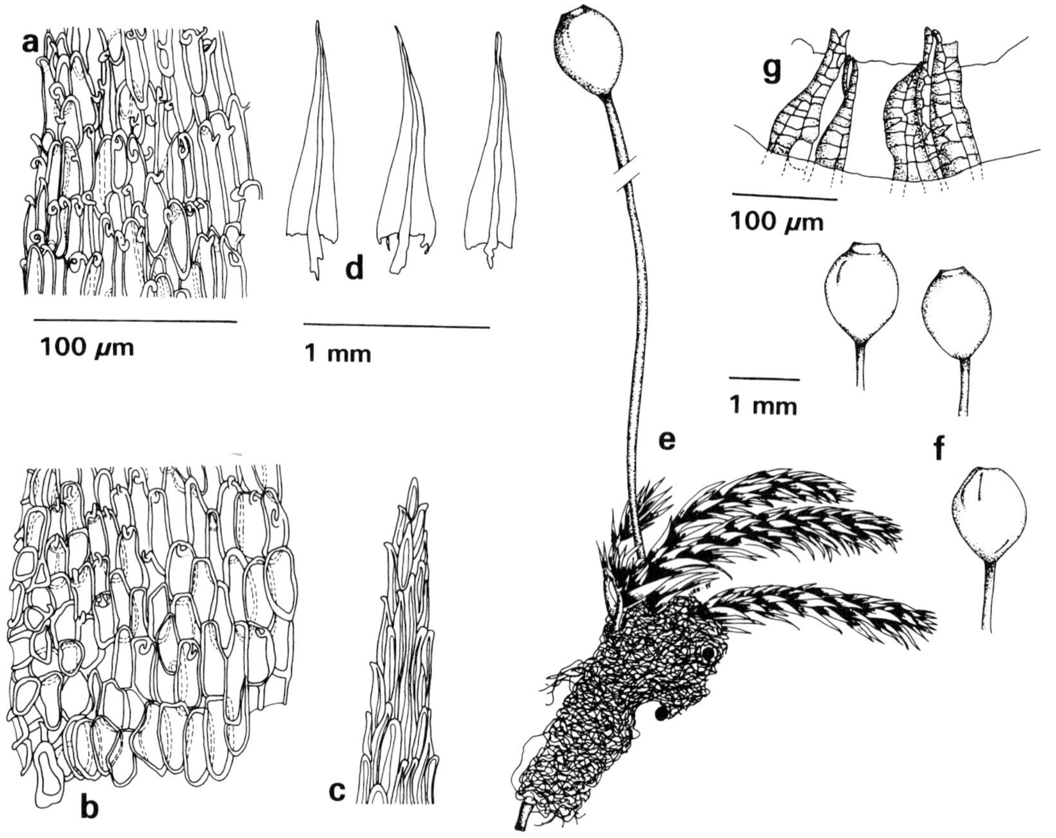

Fig. 2. *Philonotis calomicra* Broth. (from *Koponen 27756*, H). — a: Leaf cells in the upper leaf. — b: Basal leaf cells. — c: Leaf apex. — d: Leaves. — e: Growth habit. — f: Capsules. — g: Peristome, inner surface. — Use the 100 µm scale on the left for a–c, the 100 µm scale on the right for g, the 1 mm scale on the left for d, and the 1 mm scale on the left for e and f.

cies have the capsule inclined to horizontal, often gibbous and sulcate when dry, with both the exostome and endostome present. *Philonotis calomicra* is only slightly larger than *P. hastata*, but is easily separated from it by its papillose – mamillose leaf cells, the dense areolation of the distal leaf and the generally excurrent costa. The leaves of *P. runcinata* differ from those of *P. calomicra* being narrower, and having high prorations on the distal end of the narrow leaf cells which reach the leaf base. The basal marginal cells of *P. runcinata* have protruding cell corners. In contrast, *P. calomicra* has basal cells which are rectangular and broadly mammillate. Its basal marginal cells are smooth. Only rarely are the leaves in *P. calomicra* revolute - recurved, and this separates them from *P. thwaitesii*. The specimens *Koponen 27754* and *27756* in which *P. calomicra* and *P. runcinata* are mixed with each other were crucial for understanding the taxonomy of these taxa (Figs. 2 and 5).

On the basis of the characters of the capsule and peristome (endostome lacking) Brotherus (1924) placed *Philonotis calomicra* in the section *Leiocarpus* Broth. In the same section he placed two species of *Bartramidula*, *B. tjibodensis* Fleisch. and *B. treubii* Fleisch., which Fleischer (1904) described from Java. A fourth species with an erect subglobose capsule, *Philonotis imperfecta* Bartr., is known from North Borneo. Iwatsuki and Noguchi (1975) compared *Philonotis treubii* and *P. imperfecta* with each other, and found *P. treubii* to differ in having prolonged excurrent costa, larger leaves and smooth to papillose peristome teeth, while the costa in *P. imperfecta* is percurrent to slightly excurrent and peristome is smooth. Our plants from New Guinea are identical to *P. imperfecta*. The name *P. calomicra* antedates *P. imperfecta*. Our New Guinea material of *P. calomicra* differs from *P. treubii* in having the dry capsule globose while the dry capsule of *P. treubii* tends to be ellipsoid. This may be due to that the exothecial cells in *P. calomicra* are thin-walled and those in *P. treubii* have more thickened cell-walls with distinct trigones. Some specimens published as *P. tjibodensis* are actually *P. calomicra*, e.g. *Williams 2390* (H-BR) and *Ramos & Edano 28939* (H-BR) from the Philippines (Bartram 1939), and *Robbins 3029* (FH) from New Guinea (Bartram 1961).

Philonotis calomicra was collected only in two localities during the Koponen-Norris expedition in 1981. One was on gravel on a road side bank at ca. 600 m, and the other was on soil in a rivulet in a tropical rainforest at ca. 800 m. Streimann collections, which indicate that *P. calomicra* is more common in suitable lowland sites, were collected between 500–1 980 m on disturbed habitats such as road cuttings.

Range on the Huon Peninsula: *14c*. 27754 (with *P. runcinata*), 27756 (with *P. runcinata*). *14e*. 59094, 59095.

Range in Western Melanesia: Papua New Guinea. *9*. Frieda River Base Camp at Horse Creek, open gardens and lawns, on soil, *Koponen 34771, 34816, 34817* (H, as *P. revoluta* in Norris & Koponen 1985). *12*. Present report and: Lae, Institute of Technology grounds, on soil bank in open position near sea level, *Argent 27/5/71/1* (LAE); Aseki-Menyamya Road, Spreader Divide, 12 km of Aseki, *Streimann 12063, 12782* (H); Wau-Bulolo Road 11 km SW of Bulolo, on shaded roadside, 750 m, *Streimann 22868* (H). *14*. Mondomil Mission, 5 km SE of Minj, on stones on Minj River Bridge, *Streimann 21450* (H). *19*. Bema–Kaintiba Road, 4 km N of Kaintiba, *Streimann 33649* (H). *20*. Hombrom Bluff Road, 26 km NE of Port Moresby, *Eucalyptus*-dominated savannah woodland on conglomerate in seepage area, *Streimann 15007* (H).

Other specimens studied (new country records). — Philippines. *Williams 2390, Ramos & Edano 28939*, as *Philonotis tjibodensis* (H-BR, NY). Western Samoa. Upolu, Apia, *Fleischer B 1149* (NY, as *P. hastata* in Schultze-Motel 1974).

Total range: As 4: Phi PNG Sabah. Oc: Sam.

2. *Philonotis hastata* (Duby) Wijk & Marg. (Figs. 3, 8)

Taxon 8: 74. 1959. — *Hypnum hastatum* Duby in Moritzi, Syst. Verz. Zoll. Pfl. 132. 1846. — Type: Indonesia. Java, ad rupes fl. Tjappus, 1813 *Zollinger* (L 910, 104-376, not seen; cf. Iwatsuki 1977).

Bartramia tahitensis C. Müll., Bot. Zeitung (Berlin) 17: 220. 1859. ("*Tahitensis*") — *Philonotula tahitensis* (C. Müll.) Besch., Ann. Sci. Nat. Bot. sér. 7, 20: 28. 1894. — *Philonotis tahitensis* (C. Müll.) Aongstr., Öfv. K. Vet. Akad. Förh. 30(5): 120. 1873. — Type: Society Islands. "Insula Tahiti. Van den Bosch donavit" (not seen).

Bartramia laxissima C. Müll., Syn. Musc. Frond. 1: 480, 1849, nom. illeg. — *Philonotis laxissima* Mitt., J. Proc. Linn. Soc. Bot. Suppl. 1: 61. 1859, nom. illeg. — Homotypic with *P. hastata*.

Philonotis asperifolia Mitt., J. Linn. Soc. Bot. 10: 185. 1868. — Type: Samoa. Tutuila, on damp earth and on rocks by water-courses, 20–100 ft., *Powell 28* (NY-Mitten, holotype; com. Mitten, NY, FH-Fleischer, isotypes).

P. imbricatula Mitt., Proc. J. Linn. Soc. Bot. Suppl. 1: 61. 1859. — *Bartramia imbricatula* (Mitt.) C. Müll., Linnaea 36: 1869. — Type: Sri Lanka. Ceylon, *Gardner* (NY, herb. Mitten 609, lectotype, *nov.*).

P. heterophylla Mitt., J. Proc. Linn. Soc. Bot. Suppl. 1: 61. 1859, *syn. nov.* — *Bartramia heterophylla* (Mitt.) C. Müll., Linnaea 36: 12. 1869. — Types: Sri Lanka. Ceylon, Pas dum Koyle, XII.1846 *Gardner* (NY-Mitten 612, lectotype, *nov.*); Central Prov., *Thwaites 94* (NY-Mitten, syntype; on another sheet no. *94* = no. *616*, BM, see Ochi 1970, syntype), *Thwaites 97* (H-BR, BM, see Ochi 1970, syntypes).

Bartramia obtusifolia Mitt. in Seemann, Fl. Vit. 381. 1873, *syn. nov.* — *Philonotis obtusifolia* (Mitt.) Par., Index Bryol. 926. 1897. — Type: Fiji. Ysabel, *Veitch* (NY-Mitten).

Bartramia wallisii C. Müll., Linnaea 38: 554. 1874 ("*Wallisi*"). — *Philonotis wallisii* (C. Müll.) Jaeg., Ber. St. Gallischen Naturwiss. Ges. 1877–78: 437 (Gen. Sp. Musc. 1: 701). 1880. — *Fleischerobryum wallisii* (C. Müll.) Loeske, Stud. Morph. Syst. Stud. Laubm. 127. 1910. — Type: Philippines. Insulae Philippinae, Luzon, Mahahai, regione montosa, 1871 *Wallis*, "Hb. C. Müll." (H-BR, lectotype, *nov.*; BM-Bescherelle, isolectotype, BM-Hampe, isolectotype). — Synonymized by Bartram (1939).

Philonotula jardinii Besch., Ann. Sci. Nat. Bot, sér. 7, 20: 29. 1894 ("*Jardini*"). — *Philonotis jardinii* (Besch.) Par., Index Bryol. 923. 1897. — Type: Society Islands. Tahiti, 1852 ex herb. É. Jardin (herb. Bescherelle, BM, holotype). — Synonymized by Whittier (1976).

P. evaninervis Fleisch., Musci Fl. Buitenzorg 2: 612, *fig. 116*. 1904, *syn. nov.* — Type: Indonesia. Java. Bei Tjiburrum, 1 650 m, *Fleischer* (FH-Fleischer, lectotype, *nov.*); Sindanglaya an feuchten Mauern an der Strasse, 1 050 m, IV.1901 *M. Fleischer* (H-BR, syntype).

P. tildeniae Card., Josephine E. Tilden, South Pacific Plants 71B (label). 1909(?), nom. nud., *new assign.* — Original collection: Society Islands. District of Papenoo, Tahiti, X.1909 *Tilden 71B* (E).

P. St.-Johni Bartr., Occasional Papers Bernice P. Bishop Mus. 15: 334. 1940, *syn. nov.* — Type: Gambier Islands, Aukena Island, Koiovao (Koivao), moist basal cliff, alt. 15 m, V.1934 *St. John 14667* (FH-Bartram, holotype).

Plants 0.3–2.0 cm tall, pale green, in loose tufts; tomentum scarce, branched with subfloral innova-

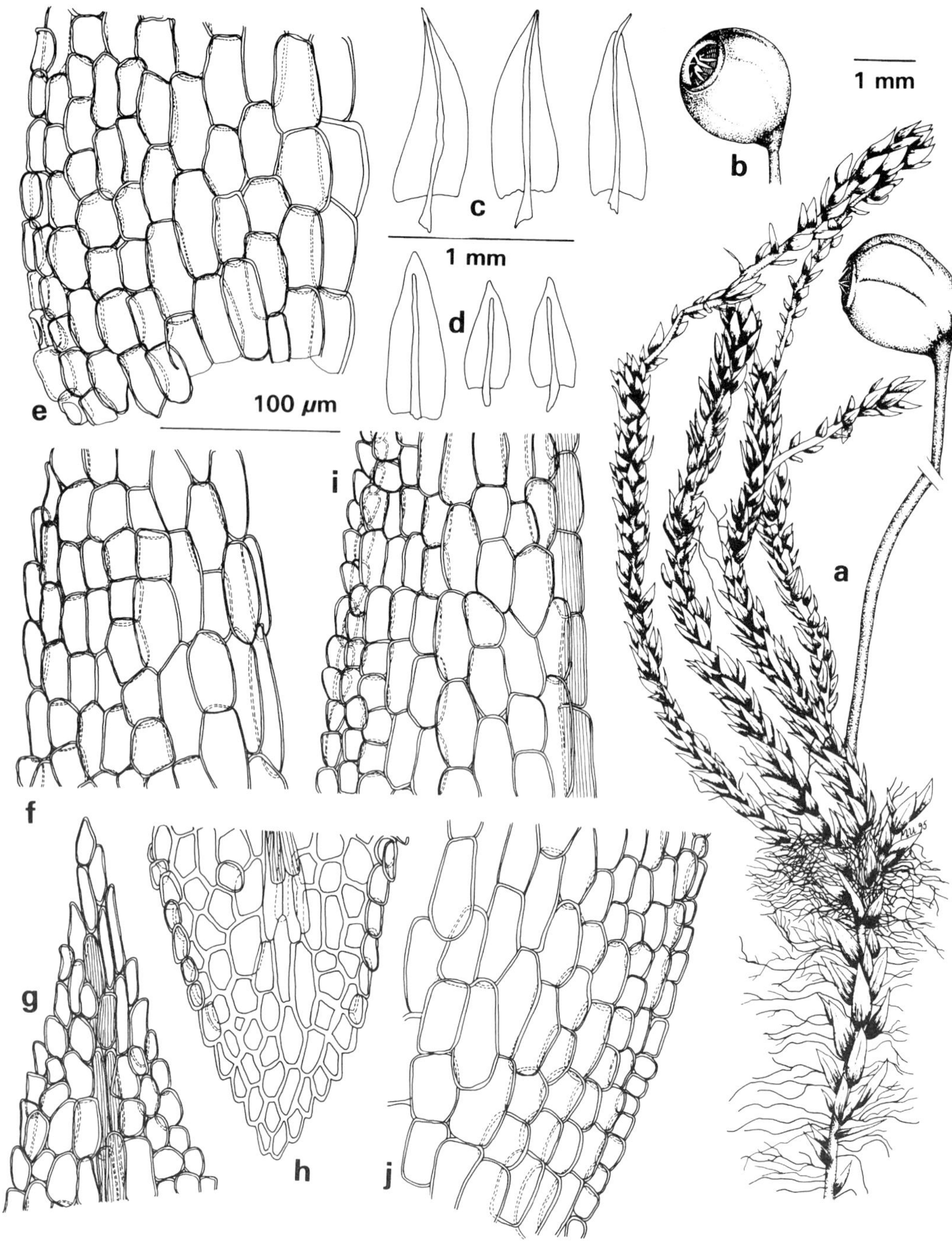

Fig. 3. *Philonotis hastata* (Duby) Wijk & Marg. (a, b, and e–j from *Koponen 29606*, H; c and d from *Koponen 29609*, H). — a: Growth habit. — b: Capsule. c: Stem leaves. — d: Branch leaves. — e: Basal leaf cells of stem leaf. — f: Distal leaf cells of stem leaf. — g: Apex of stem leaf. — h: Apex of branch leaf. — i: Distal leaf cells of branch leaf. — j: Leaf base of branch leaf. — Use the 100 μm scale for the cellular details, the 1 mm scale on the left for the leaves, and the 1 mm scale on the right for the growth habit and b.

tions, branches erect; stem cross-section with large-celled epidermis, outer cell walls (hyalodermis) thin, substereidal cortical cells in one layer, medullar and central cells thin-walled; leaves erect to patent when dry, erect spreading when moist, keeled or plane, straight or slightly curved, 0.1–0.3 × 0.6–1.0 mm, acute from an ovate base, usually broadest above base and slightly narrowed at the insertion, apex acute or obtuse, often ending with one uniseriate cell; border plane or recurved, leaf margin entire at base, mammillosely geminate towards the apex; costa weak, greenish, ca 25 µm at base, mammillosely toothed on both sides, weak at apex, usually not reaching the apex in sterile plants and on branch leaves but often percurrent but rarely excurrent on leaves from fertile stems; epidermis in cross-section thin-walled papillose-mammillose, cells at midleaf rectangular to rhomboidal, 12–22 × 25–62 µm, thin-walled, indistinctly mammillose; apical cells thin-walled, shorter, rhomboidal to subquadrate, basal cells similar to median leaf cells, thin-walled, 15–27 × 25–55 µm, often quadrate near border, rectangular near costa; fragile, leafy propagule-like branchlets are commonly present.

Dioicous. Perichaetia terminal, inner perichaetial leaves with narrow ovate base, tapering gradually to long uniseriate apex; perigonia gemmiform, perigonial leaves ca. 1.0 mm long, base of inner leaves cochleariform, apex abrubtly contracted to acute-acuminate apex; seta 1.5–2.5 cm long, capsule horizontal, arcuate or gibbose when dry, globose to short ellipsoid when wet, smooth or striate, 1.5–2 mm in diameter, mouth small, operculum low conic; peristome teeth brown, 0.3 mm long, trabeculate, papillose; endostome as long as the exostome, basal membrane half of endostome, cilia lacking or very rudimentary. Spores ca. 12–18 µm, papillose in light microscope.

Illustrations: Dozy & Molkenboer 1855–70: tab. CXXIV, as *Philonotis laxissima*; Fleischer 1904: 613 (fig. 116, as *P. evaninervis)*; Bartram 1933: 138 (fig. 101, as *P. laxissima*); Kabiersch 1938: 103 (Abb. 33, fig. 2–4); Bartram 1939: pl. 12 (fig. 198, as *P. imbricatula*, fig. 199, as *P. laxissima*); Ochi 1962: Tab. 40 (fig. A–C, as *P. evaninervis*, D–G); Ochi 1970: 24 (fig. 48, as *P. heterophylla*); Whittier 1976: 189 (fig. 52 D–G); Iwatsuki 1977: 14 (fig. a–o, type specimen); Noguchi 1989: 573 (fig. 254A, as *P. evanidinervis*, fig. 254B).

Whittier (1976) synonymized *Philonotula jardinii* on the basis of Bartram's (1931) comment that there is little or nothing in the description to distinguish it from *Philonotis laxissima*. Neither of them saw the type.

The specimen selected as the lectotype of *Philonotis evaninervis* bears the mark *n. sp.* by Fleischer's hand.

There seems to be some confusion with the syntypes of *Philonotis imbricatula* in the Mitten herbarium. There are two different numbering systems on the specimens. Mitten has his own numbering running more or less regularly through his publication (1859) from no. 1 (*Andreaea rigida*) to no. 1282 (*Sphagnum khasianum*). The second numbering of the Ceylon Mosses (abbreviated as C. M. on the labels of different sets in Mitten's and other herbaria, at least in BM, H-BR and H-SOL) are Thwaites' collection numbers. Three collection numbers (Mitten 1859) are given as syntypes: *509* (= Ceylon Mosses 67), selected here as the lectotype, *613* (= C. M. 91) and *616* (= C. M. 94). The specimen no. 509 fits the description better than the previous two. We identified Mitten's no. 613 as *P. thwaitesii*, and this is confirmed by Ochi (1970) who gave "Thwaites CM 91" as the isotype of that taxon. Mitten (1859) not only gave the number 616 as the type of *P. imbricatula* but also as the type of *P. heterophylla* in the same publication "*Hab.* in Ceylon, *Gardner* (No. 616) *et Thwaites* ?" (and as one of the collection numbers under *P. turneriana* (Schwaegr.) Mitt.). Ochi (1970) studied "Thwaites CM 94 & 97" in BM and gave them as the syntypes of *P. heterophylla* Mitt. and explained that they are different from *P. hastata* (syn. *P imbricatula*). We have studied those specimens (CM 94 in NY, herb. Mitten, CM 97 in H-SOL) and we do not agree with Ochi's conclusion (see below).

Ochi (1970) synonymized *Philonotis asperifolia* with *P. heterophylla* from Ceylon. We agree and believe that both are conspecific with *P. hastata*.

Philonotis hastata is easily separated by its small leaves which are only 1 mm, by its wide isodiametric, long rectangular or rhomboidal leaf cells and by a weak costa, which may be percurrent only on stem leaves of fertile plants. In sterile plants and on branch leaves of fertile plants the costa does not reach the apex and there are usually several wide thin-walled cells between the end of costa and the leaf apex. Even in plants with a seemingly percurrent costa the costa is apically weak and rarely excurrent in a stiff point as in most New Guinea species of the genus. The leaf cells are slightly mammillose while the leaf cells of all other New Guinea species of *Philonotis* are prorate or high papillose at least in the apical portion of the leaf. The teeth of the leaf border in *P. hastata* are blunt, while in the other species they have sharp teeth. The hyalodermis of *P. hastata* is formed by epidermal cells larger than the substereidal cells. In the other species treated here the hyalodermis cells are of a small size. Another anatomical difference seems to be the lack of specialised stereidal cells in the costa. The spores of *P. hastata*, 12–18 μm are smaller than those of other New Guinea species (ca 25 μm) of *Philonotis*.

The only other New Guinea *Philonotis* with similarly lax areolation is *P. vescoana*. The short-leaved forms have rather wide leaf cells at the apex, but they are distinctly papillose-mammillose, and the costa of *P. vescoana* is always very strong and excurrent. The plants from various parts of the Pacific, India and SE Asia known as *P. imbricatula* have percurrent or slightly stronger costa than that described above for the New Guinea plants. The leaf apex is also acute and the apical leaf cells are narrower. Ochi (1970) kept these plants separate under the name *P. heterophylla*. The stem leaves in the type of *P. heterophylla* have a strong costa, but the branch leaves are very similar to those of *P. hastata*. We could not find any constant characters separating these populations and therefore synonymize them. Bartram (1939) was of the same opinion when he synonymized *P. wallisii* with *P. laxissima* (= *P. hastata*).

Most plants of *Philonotis hastata* have acutish leaves which end with an uniseriate cell. However, the leaf apex may also be more or less rounded. This may be seen on a single stem where the leaves at the base of shoot are more broadly acute than those at the apex. The extreme morph in this respect are plants named as *P. evaninervis* Fleisch. We do not think that it deserves any taxonomic status and thus synonymize it.

Philonotis calomicra is nearly as small as *P. hastata*, but it has clearly papillose cells, even at the basal of the leaf, and the costa is excurrent.

The altitudinal range of *Philonotis hastata* is from sea level to 2 000 m. Original habitats appear to be river shore cliffs and banks. Most of our specimens were collected from rather open disturbed sites, such as gardens and villages. The substrates were: on bank or soil (6 specimens), cliff and boulder (2) and tree trunks (3). Eleven out of the 13 New Guinea specimens studied were taken from near rivers or streams.

Range on the Huon Peninsula: *3c*. 29538, 29606, 29650. *3i*. 60467. *3l*. 61091. *4d*. 66401. *6p*. 33019. *13d*. 62165, 62178, 62179, 62186. *16b*. 59119.

Range in Western Melanesia: West Irian. *7*. Zanten (1964, as *Philonotis evanidinervis*). Papua New Guinea. *12*. Present report and: Lae, Botanic Garden, on *Frangipani* trunk in garden, *Argent 4* (LAE). *13*. Bartram (1961a, as *P. laxissima*). *19*. Werr River, Kaintiba, regrowth on old garden site beside stream, on rock, *Streimann 33568* (H). *20*. Ruburu Village, Musgrave River, 20 km E of Sogeri, poor vegetation on steep rocky river bank, on moist shaded rock, 400 m, *Streimann & Naoni 15135* (H). *27*. Tinputs District, Hakau Plantation, lowland rainforest, on rock near slow running creek, 5 m, *Korea 4* (H). Solomon Islands. *33*. Puepue Creek above Kira-Kira, in moderately disturbed lowland forest, moist, sunny soil of riverbank, 0–60 m, *Norris 49107 & Roberts* (H); Maghoha River near Manitatama, low elevation rainforest, on sand under boulder at river, 130–140 m, *Norris 48883 & Roberts* (H).

Total range: Eur: Spain (Frahm 1976). Widely distributed in SE Asia and the Pacific. Confirmed specimens for: As 2: Chi. As 3: Sri. As 4: Ind (Java, West Irian) Phi PNG. Oc: Fij NC Sam Soc Van. Austr 1.

3. *Philonotis mollis* (Dozy & Molk.) Mitt. (Figs. 4, 9)

J. Linn. Soc. Bot. Suppl. 1: 60. 1859. — *Bartramia mollis* Dozy & Molk., Ann. Sci. Nat. Bot., ser. 3, 2: 300. 1844. — Type: Indonesia. Java, Sallak, *Kühl & van Hasselt* (L, lectotype, *nov*).

Bartramia secunda Dozy & Molk., Plant. Jungh. 3: 332. 1854 — *Philonotis secunda* (Dozy & Molk.) Bosch & Sande Lac., Bryol. Jav. 1: 156, tab 126. 1861. — Type: Indonesia. "Habitat insulam Javae", *Junghuhn* (not seen, two possible syntypes in H-SOL).

Philonotis mollis (Dozy & Molk.) Mitt. var. *simplicicaulis* Zanten, Nova Guinea Bot. 16: 293, *pl. 25 fig. 1*. 1964, *syn. nov*. — Type: Indonesia. West Irian, Jayawijaya, Star Mts, Mt Antares, Camp 39a, on trunk of tree in rain forest, alt. 1 500 m, 11. VII. 1959 *van Zanten 496b* (L, holotype; syntypes nos. *498b* and *874b* in L).

Plants to 1 cm tall, pale green, in loose tufts; stems kept together by dense tomentum, branched with subfloral innovations, branches flexuose; epidermis of stem cross-section with thin outer cell walls (hyalodermis), one layer of substereidal cortical cells and thin-walled medullary and central cells; leaves soft, erect spreading when dry, erect spreading to patent when moist, leaves on sterile plants spreading and distant, smooth, flexuose, 1–1.5 mm, linear-lanceolate, apex piliferous; border plane, basal marginal cells entire, geminate towards apex; costa brownish, 25 μm at base, nearly smooth on ventral side, excurrent, epidermal cross-section smooth, few conductive cells and dorsal stereidal cells; cells at mid-leaf rectangular to linear, 7–10 × 30–70 μm, thin-walled or with slightly thickened walls with mammilla on the ventral upper cell ends, distal cells narrower than

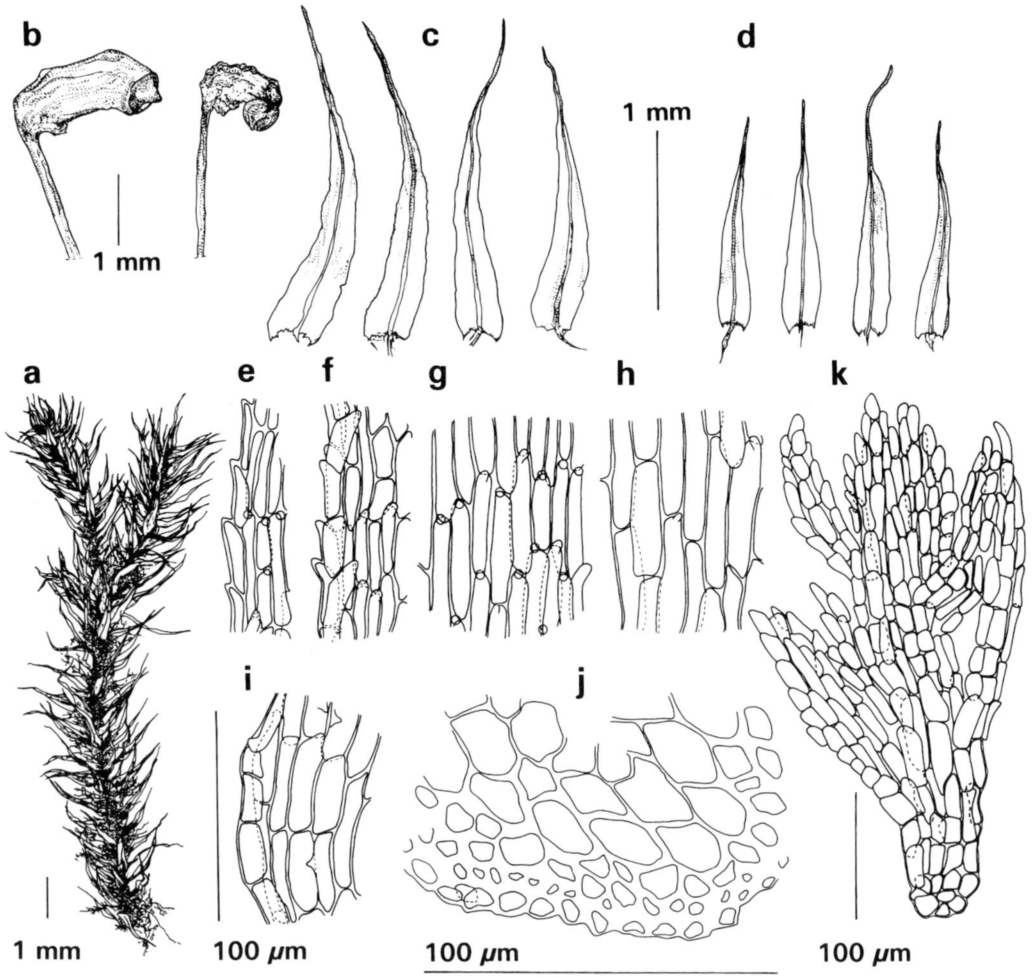

Fig. 4. *Philonotis mollis* (Dozy & Molk.) Mitt. (a, c–k from *Koponen 32138*, H; b from *Streimann 25592*, H). — a: Growth habit. — b: Wetted and dry capsule. — c: Stem leaves. — d: Branch leaves. — e: Distal leaf border. — f: Mid-leaf border. — g: Mid-leaf cells. — h: Basal leaf cells. — i: Basal leaf border. — j: Cross-section of stem. — k: Leafy propagule. — Use the 1 mm scale in lower left corner for a, the 1 mm scale in upper left corner for b, the 1 mm scale on the right for c and d, the 100 µm scale on the left for e–i, the 100 µm scale in the middle for j, and the 100 µm scale on the right for k.

mid-leaf cells, mammillose/papillose, basal cells wider and longer, thin-walled, 7–16 × 30–80 µm, elongate; small leafy axillary propagules often present.

Dioicous. Perichaetia terminal, inner perichaetial leaves from ovate base abruptly contracted to filiform apices; perigonia gemmiform, leaves 2 mm long, inner leaves with brownish, cochleariform base, abrubtly contracted to acute-acuminate apex; only one capsule seen in New Guinea specimens; but is horizontal, striate, pale and glossy.

Illustrations: Dozy & Molkenboer 1845–48: tab. CXXV, CXXVI as *Philonotis secunda*; Fleischer 1904: 617 (fig. 117); Bartram 1939: pl. 12 (fig. 202); Gangulee 1974: 1121 (fig. 545); Noguchi 1989: pl. 18 (fig. 251B).

Major taxonomic problems are concentrated in taxa known earlier as *Philonotis secunda* and *P. revoluta*, as well as the only recently recognized and widely distributed *P. runcinata* and *P. vescoana*. *Philonotis secunda* was originally described as *Bartramia secunda* by Dozy and Molkenboer (1854) and transferred to *Philonotis* by Bosch and Sande Lacoste (1861), who at the same time retained *Bartramia mollis* Dozy & Molkenboer (1844), and described some new taxa. One of them was *P. revoluta*. The plates 126 and 128 in Bosch and Sande Lacoste (1861) show two very similar plants. The major differences are that *P. secunda* has leaves with longer excurrent costa and the leaf hair-point is more or less flexuose. Also, *P. secunda* has wider and longer basal cells which actually is the diagnostic character of *P. mollis*. The inner perigonial leaves in *P. secunda* are abruptly acuminate while those in *P. revoluta* are more gradually tapering. Fleischer (1904) retained both species on the basis of the difference in the perigonial leaves. He added that the ordinary leaves are longer and with less recurved border in *P. secunda*. Kabiersch (1938) cited both taxa and stated that *P. secunda* differs from *P. revoluta* by one-sidedly secund and narrower leaves and by the wider areolation of the cells. He added that the difference in the length of the setae used by Brotherus (1924) is not useful. Ochi (1962, 1963) did not deal with *P. secunda* and, according to him, the records for Japan are based on erroneous determinations. Bartram (1939) recognizes both species and he separated them in his key on the basis that *P. secunda* has leaf margins which are slightly revolute, and elongate leaf cells that are narrowly rectangular or linear, while in *P. revoluta* the leaf margins are strongly revolute, and has shorter leaf cells that are more uniformly rectangular. Hegewald and Zanten (1986) mentioned that the differences between *P. nitida* and *P. secunda* are slight and cite the Fleischer characters of the perigonial leaves as diagnostic characters. They gave *P. nitida* as a synonym of *P. revoluta* (see under *P. thwaitesii*).

The solution of this problem is that *Philonotis revoluta* is actually a synonym of *P. thwaitesii*, as is discussed below, and *P. secunda* may be identical with *P. mollis*. This does not, however, solve all the problems; we have seen specimens named as *P. revoluta* from the Philippines and Melanesia which do not belong to any of the taxa occurring in New Guinea. They maybe *P. turneriana*, but this must be confirmed.

Philonotis mollis has laxer areolation in the basal leaves than any other *Philonotis* in our study area, except *P. hastata*. Their leaf shapes, however, are quite different, *P. mollis* having long and narrow, gradually tapering leaves, while the leaves in *P. hastata* are ovate. Also, the long and narrow basal leaf cells separate *P. mollis* from the other species. The soft texture of the spreading and distant leaves on lower and sterile stems are among the diagnostic characters of *P. mollis*, as well as the shiny capsule.

Very slender shaded forms of *Philonotis mollis*, e.g. those named as *Philonotis mollis* var. *simplicicaulis* (see above) grow often mixed with other mosses on soil or epiphytic, and may be difficult to separate from *P. runcinata*. These shaded forms are deep green in colour while *P. runcinata* is a pale colour. The other differences between *P. mollis* and *P. runcinata* are discussed under the latter name.

Philonotis mollis grows between 800–3 250 m both in primeval and disturbed forests. Most of the specimens were collected from soil along trails (3 specimens), and one specimen on rotten log, and cliff.

Range on the Huon Peninsula: *1d.* 34749. *4g.* 66665. *6d.* 32138. *10a.* 28301. *14e.* 27939, 27995B.

Range in Western Melanesia: West Irian. *7.* Zanten (1964). Papua New Guinea. *12.* Present report. *24.* West New Britain, 25 km SE of Kimbe, oil palm plantation, 50 m, *Suarim 16* (H).

Total range: As 2: Chi Ja; As 3: In; As 4: Ind (Java, West Irian) Phi PNG.

4. *Philonotis runcinata* C. Müll. ex Aongstr. (Figs. 5, 9)

Öfvers. Kongl. Svenska Vetensk. -Akad. Förh. 33(4): 52. 1876. — *Philonotula runcinata* (Aongstr.) Besch., Ann. Sci. Nat. Bot. sér. 7, 20:28. 1894. — Type: Society Islands. "Tahiti", *Andersson* (B-BR, H-SOL, isotypes).

Philonotis praemollis Broth. & Par., Öfvers. Förh. Finska Vetensk.-Soc. 51 A(17): 20. 1909, *syn. nov.* — Type: New Caledonia. Mont Dzumac, ad terram, VII.1908 *Le Rat* (H-BR, lectotype, *nov.*).

P. parisii Thér., Diagn. Esp. Var. Nouv. Mouss. 8: 5. 1910 ("*Parisi*"), *syn. nov.* — Type: New Caledonia. Col d'Annieu, VIII.1908 *Marc Buso 1196* (H-BR, isotype).

P. angustissima C. Müll. ex Tixier, Cryptog. Bryol. Lichénol. 7: 228, *fig. 2.* 1986, *syn. nov.* — *P. angustissima* (C. Müll.) Par., Index Bryol. 917. 1897, nom. inval. — *Bartramia angustissima* C. Müll., Rev. Bryol. 14: 57. 1887, nom. nud. — Type: New Caledonia. Mont Atso, *Saves 18* (H-BR, isotype).

P. yunckereana Bartr. in Yuncker, Bernice P. Bishop Museum Bulletin 220: 15. 1959, *syn. nov.* — Type: Tonga. Niuafo'ou, on soil in open area just above 1946 lava flow, on way to summit of main crater, 16.V.1953 *Yuncker 16017* (FH-Bartram, holotype, NY, 2 isotypes).

Plants ca. 1 cm tall, pale green, in loose tufts; stems kept together by dense tomentum, branched with subfloral innovations, branches flexuose; epidermal stem cross-section with thin outer cell walls

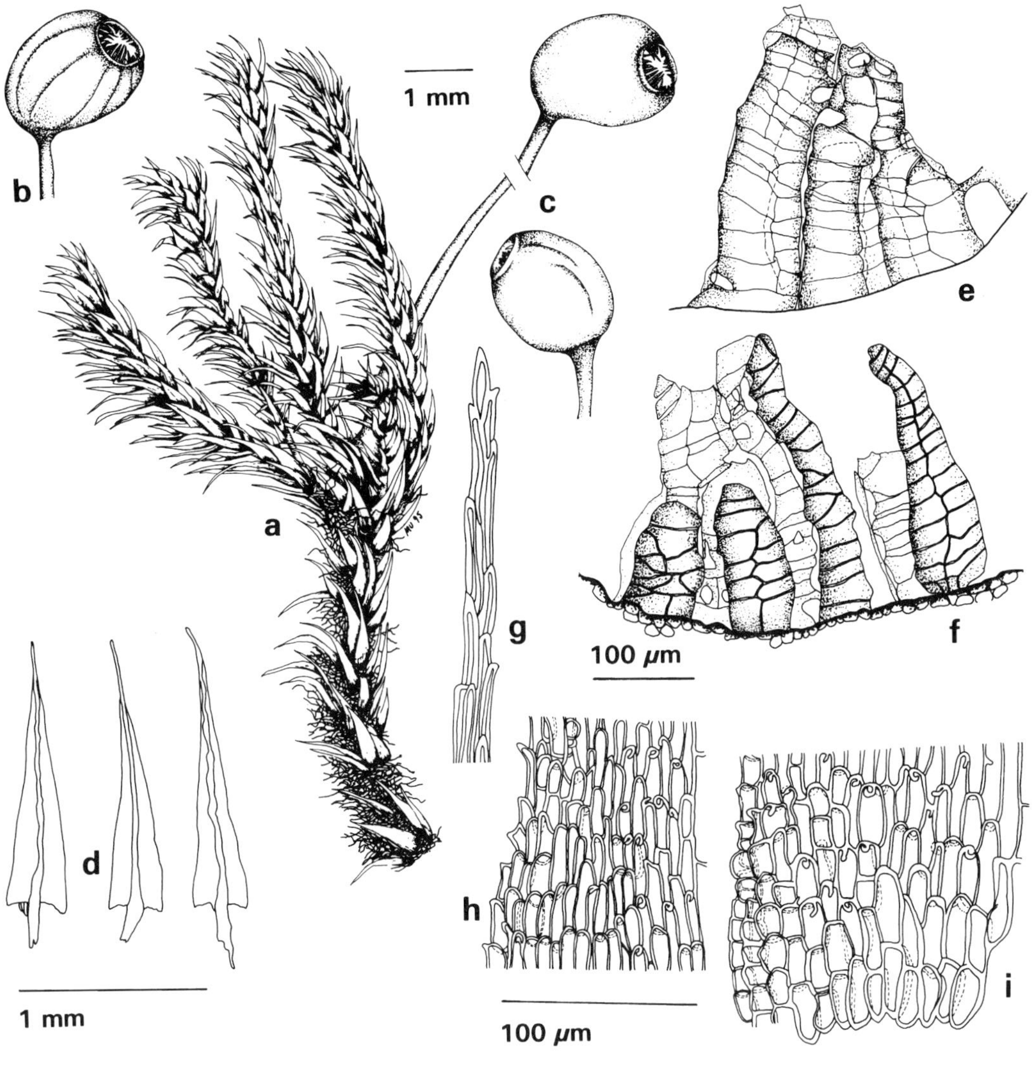

Fig. 5. *Philonotis runcinata* C. Müll. ex Aongstr. (from *Koponen 27756*, H). — a: Growth habit. — b, c: Capsules. — d: Leaves. — e: Endostome, inside view. — f: Peristome and endostome, outside view. — g: Leaf apex. — h: Distal leaf cells. — i: Leaf base. — Use the 100 µm scale below for the cellular details, the 100 µm scale above for the peristomes, the 1 mm scale on the left for leaves and the 1 mm scale on the right for the growth habit and capsules.

(hyalodermis), one layer of substereidal cortical cells and thin-walled medullary and central cells; leaves erect spreading when dry, patent to spreading and secund when moist, plane, falcate, 0.2–0.4 × 0.8–1.2 mm, acuminate from narrowly ovate base, broadest at insertion or just above it, apex acute, not piliferous; border plane or revolute, basal leaf margin usually crenulate by protruding cell corners, geminate-denticulate towards apex; costa brownish, 20–30 µm at base, papillosely toothed on ventral side, percurrent or shortly excurrent, cross-section with papillose-mammillose epidermis, few conductive cells, other cells thick-walled; midleaf cells rectangular to linear, 5 × 25–37 µm, thin-walled with projecting papilla at upper end of cell, apical cells similar to midleaf cells, basal cells wider and shorter, thin-walled, 10 × 12–35 µm, often quadrate near border, rectangular near costa.

Dioicous. Perichaetia terminal, inner perichaetial leaves with filiform apices; perigonia gemmiform, leaves ca. 1.5 mm long, base of inner leaves cochleariform, abrubtly contracted to an acute-acuminate apex; seta 2–3 cm long, capsule horizontal to inclined, globose to ellipsoid, sulcate when dry, ca. 1–1.5 × 1.5–2 mm, constricted below mouth when dry; operculum low conic; peristome teeth brown, 0.4 mm long, papillose; endostome nearly as long as exostome, basal membrane half of endostome, cilia lacking. Spores ca. 25 µm, papillose in light microscope.

Illustrations: Tixier 1986: 229, (fig. 2, as *Philonotis angustissima*); Whittier 1976: 189 (fig. 52 H-K).

The growth habit characteristics of *Philonotis runcinata* are the small size, the rather pale colour, the narrow and secund to flexuose but imbricate leaves when dry, which are erect spreading to patent when moist. If no sporophytes are present then it can be confused with *P. calomicra* which is of the same size. Microscopic characters which separate *P. runcinata* and *P. calomicra* are differences in the leaf shapes, areolation and mammillosity/papillosity of leaf cells, and in marginal teeth of the leaves. The leaves of *P. runcinata* are broadest at insertion, *P. calomicra* has shortly tapering leaf bases. In *P. runcinata* the basal marginal cells are quadrate and crenulate and the basal leaf cells are distinctly papillose on distal cell ends, while in *P. calomicra* the row of basal marginal cells is entire and the basal cells bave more distinct mammilla on the proximal cell end. The papillae of *P. calomicra* are not quite as high as in *P. runcinata*. Small plants of *P. mollis* may resemble *P. runcinata*. Both have a similar wirey appearance when dry under the binocular microscope; they have rather similar leaf areolation and serrations; and both species have leaves broadest at insertion or just above it. The basal leaf cells of *P. mollis* are comparatively broader and longer, and less papillose which gives the basal areolation a laxer appearance than in *P. runcinata*. It differs from *P. thwaitesii* by its much narrower leaves with non-recurved borders, or only one leaf border being recurved

Philonotis runcinata has a rather wide altitudinal range in New Guinea. Our Huon Peninsula specimens were collected from near sea level to 3 600 m on moist soil, gravel, cliffs or decaying wood at streams or trails. The Streimann collections (14 specimens) were collected from similar habitats at altitudes between 40 and 2 000 m.

Range on the Huon Peninsula: *2b*. 33414. *2p*. 34423, 34425. *4e*. 66497. *3f*. 29785. *6r*. 33175. *14c*. 27760, 27754 (with *P. calomicra*), 27756 (with *P. calomicra*), 58878. *14d*. 58780. *15a*. 28116.

Range in Western Melanesia (first records): West Irian. *4*. *Van Zanten 193* (NY), *891b* (L, NY, as *P. revoluta* in Zanten 1964). Papua New Guinea. *10*. Kairuru Island, hill forest, on ground, 600 m, *Borrell 81-22* (H). *11*. Manam Island, Waris, talus vertical humide au bord du sentier du village, 40 m, *De Sloover 42652* (H), Bodua, près de la plage, sur le sol, *Iserentant 9395* (H). *12*. Present report and: Gabensis Creek, 30 km W of Lae, shaded section of exposed cliffs beside stream, 40 m, *Streimann & Umba 11193* (H), Upper Watut River, 9 km SW of Bulolo, on moist road cutting, shaded by grasses, 900 m, *Streimann & Kairo 17215* (H). *20*. Angabanga River, 39 km ENE of Bereina, regrowth beside road, moist, shaded road cutting, 650 m, *Streimann & Naoni 16255* (H). *24*. West New Britain, Galai, 30 km SE of Kimbe, on road side, 60 m, *Kolema 52, 62* (H).

Other specimens studied (new country records). — Malaysia. Sabah, North Borneo, West Coast. Res., Bt. Kulong near Ranau, 1 500 ft, on serpentine at margin of forest and cattle grazing area, 1961 *Meijer B 12737* (NY). Philippines. Negros Orientale, near Dumaguete, 1951 *Mack 1370* (NY). Fiji. Viti Levu, Ba Prov., Mt. Lomalagi E of Nadarivatu, on moist soil on trail in dense rainforest, 800–900 m, *Norris 44500* (H).

Total range: Probably widely distributed in the Pacific area. Confirmed for: As 4: Ind (West Irian) Phi PNG. Sabah. Oc: Fij NC Soc.

5. *Philonotis thwaitesii* Mitt. (Figs. 6, 9)

Proc. J. Linn. Soc. Bot. Suppl. 1: 60. 1859 ("*Thwaitesii*"). — Type: Sri Lanka. Central Prov. Ceylon, *Thwaites C.M. 91* (NY herb. Mitten, lectotype, *nov.* ; NY herb. Mitten, H-SOL, isolectotypes).

P. revoluta Bosch & Sande Lac., Bryol. Jav. 1: 158, *pl. 128*. 1861, *syn. nov.* — Type: Indonesia. Java, *Junghuhn* (L, herb. Sande Lacoste, lectotype, *nov.*).

P. socia Mitt., J. Linn. Soc., Bot. 8: 151. 1864. — Type: Japan. Nagasaki, *Oldham* (H-SOL). — Synonymized by Ochi (1970).

Plants ca. 1 cm tall, pale or bluish, in dense tufts; stems kept together by tomentum, branched with subfloral innovations, branches stiff; epidermis of stem cross-section with thin outer cell walls (hyalodermis), one or two layers of substereidal cortical cells and thin-walled medullary and central cells; leaves erect or falcate, imbricate when dry, erect to erecto-patent when moist, smooth, slightly falcate or straight, 0.4–0.5 × 1.6–2.0 mm, acuminate from narrowly ovate base, broadest at insertion or just

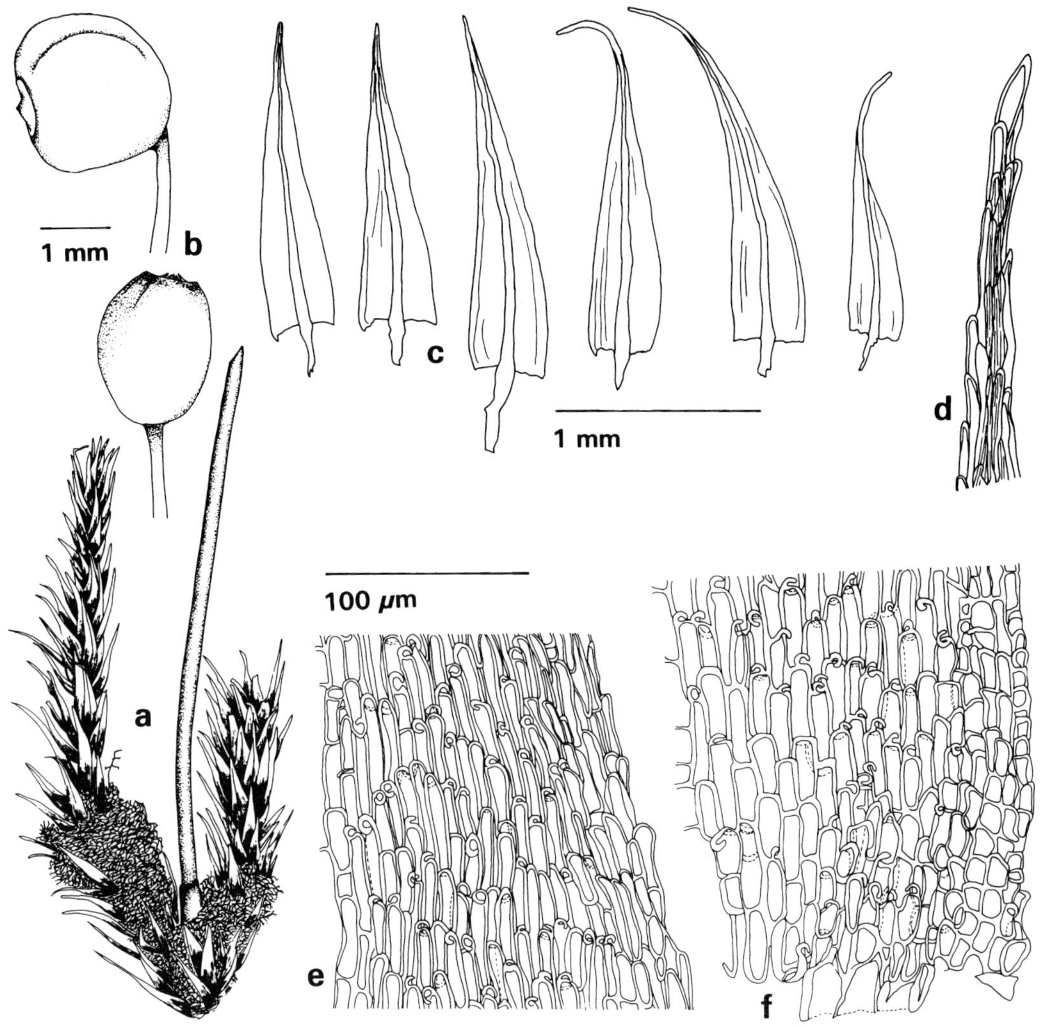

Fig. 6. *Philonotis thwaitesii* Mitt. (from *Norris 62633*, H). — a: Growth habit. — b: Capsules. — c: Leaves. — d: Leaf apex. — e: Distal leaf cells. — f: Leaf base. — Use the 100 μm scale for the cellular details, the 1 mm scale on the left for the growth habit and capsules, and the 1 mm scale on the right for the leaves.

above it, apex stiff; border recurved, revolute or plane in small leaves, leaf margin crenulate by protruding cell corners at base, geminate-serrulate towards apex; costa brownish, 55–65 μm at base, papillose toothed on ventral side, excurrent into the stiff point, epidermis in cross-section papillose-mammillose, few conductive cells and dorsal stereidal cells; cells at midleaf rectangular to linear, 5 × 20–32 μm, thin-walled or with slightly thickened walls with projecting papilla on ventral upper cell ends, cells at apical leaf as wide as mid-leaf cells, basal cells wider and shorter, thin-walled, 7–12 × 15–25 μm, quadrate to shortly rectangular.

Dioicous. Perichaetia terminal, inner leaves from ovate base abruptly contracted to filiform apices; perigonia gemmiform, leaves 1.4–2.5 mm long, base of inner leaves cochleariform, apex abrubtly contracted to an acute-acuminate apex; seta 1.5–3.5 cm long, capsule horizontal to inclined, sulcate when dry, globose when wet, ca. 2 mm long, constricted below mouth when dry; operculum low conic; peristome teeth brown, 0.6 mm long, papillose; endostome pale brown, as long as exostome, basal membrane two thirds of endostome, few very reduced cilia present. Spores ca. 25 μm

Illustrations: Dozy & Molkenboer 1855–70: tab. CXXVIII (as *Philonotis revoluta*); Brotherus 1924: 463 (fig. 405 A-D); Kabiersch 1938: 98 (Abb. 31, fig. 1–3, as *P. socia*); Bartram 1939: pl. 12, fig. 201, as *P. revoluta*); Ochi 1963: tab. 9 (7) (fig. A–S, as *P. socia*); Ochi 1970: 25 (fig. 49); Gangulee 1974: 1122 (fig. 546); Iwatsuki & Mizutani 1972: pl. 18 (fig. 252 as *P. socia*); Noguchi 1989: 561 (fig. 245 B).

In Mitten's herbarium there are two specimens of "Thwaites Ceylon Musci" no. *91*. The specimen with the text "*B.* [*artramia*] *P.* [*hilonotis*] *Thwaitesii* M.[ihi] M[usci] In[diae] Or[ientalis], Ceylon *Thwaites 91*." is selected as the lectotype. Another specimen "*613 Bartramia* Ceylon *91*" is on the *Philonotis imbricatula* sheet (see the discussion under *P. hastata*).

The comparison of the type specimens of *Philonotis revoluta*, *P. thwaitesii* and *P. socia* showed that they belong to the same widely distributed southeast Asian taxon. Our specimens fit wery well with these types. In the type specimen of *P. revoluta* there are no earlier annotation labels which may mean that no one studied that specimen.

Gangulee (1974) gave *Philonotis revoluta* as a synonym of *P. nitida* Mitt. We studied the type of *P. nitida* (In Himalayae boreali-occident. reg. temp., *Royle et T. Thomson 561*, NY-Mitten) and cannot agree with Gangulee's opinion. On the basis of the lanceolate leaves and long hairpoint of the leaves *P. nitida* is closer to *P. turneriana* (Schwaegr.) Mitt. We have not seen specimens fitting the type of *P. nitida* from the Pacific and therefore it may only be a continental Asiatic taxon.

The characteristics of *Philonotis thwaitesii* are the rather stiff habit, the strong costa penetrating into the stiff acumen, the comparatively narrow areolation in the distal half of the leaf, the tall papillae on cells and leaf borders which are recurved to revolute down to the leaf base. This combination of characters makes the areolation appear to be very dense when compared with the more lax areolation of e.g. *P. vescoana* or *P. mollis*. There is some variation in the cell details. The basal cells are regularly of the same width and twice as wide as the cells at mid leaf, and often quadrate. This area of ± quadrate cells looks more translucent than the denser upper leaf, and was used as the diagnostic character by Ochi (1963) for *P. socia* (= *P. thwaitesii*). The recurvature of the leaf border does not vary in *P. thwaitesii*. Both leaf borders of stem leaves are recurved or revolute, only in lanceolate leaves from branches only one of the borders may be recurved. *P. runcinata* is smaller than *P. thwaitesii*, and its leaves are much narrower at base and the leaf borders are plane or the border is revolute only on the distal part of the leaves. The differences between *P. thwaitesii* and *P. mollis* are discussed under the latter name.

Dixon (1942, 1943) first recorded a New Guinea specimen of *Philonotis thwaitesii* (*Carr 12399*, BM, NY) and Ochi (1970) reported the same specimen. We think that the specimen represents *P. hastata* with exceptionally strong costa on stem leaves. The leaf cells are only mammillose and without sickle-like prorations.

Our specimens were collected on open grasslands between 2 280–3 400 m on rocky habitats such as basic or limestone cliffs. Two of the Streimann specimens (see below) are from lower elevations and from disturbed habitats.

Range on the Huon Peninsula: *2a*. 33623. *6a*. 62313, 62360. *6b*. 62633, 62634. *6d*. 32165. *6e*. 32287. *6t*. 33260.

Range in Western Melanesia (only confirmed records given): West Irian. *7*. *Van Zanten 193* (as *Philonotis revoluta* in Zanten 1964). Papua New Guinea. *12*. Present report, and: Upper Baiune Power Station 12 km NE of Bulolo, on rocky road cutting, 780 m, *Streimann 17786* (H).

Total range: As 2: Widely distributed in Japan, Korea and Taiwan (Ochi 1963, as *Philonotis socia*). Confirmed specimens from: As 3: Sri. As 4: Ind (Java, West Irian) PNG. Oc: Ton.

6. *Philonotis vescoana* (Besch.) Par. (Figs. 7, 9)

Ind. Bryol. 931. 1897. — *Philonotula vescoana* Besch., Ann. Sci. Nat. Bot. sér. 7, 20: 29. 1894 ("*Vescoana*"). — Type: Society Islands. Tahiti, *Vesco* (herb. É. Bescherelle, BM, holotype; FH, isotype).

Philonotis eurybrochis Ren. & Card., Rev. Bryol. 23: 101. 1896, *syn. nov.* — *Fleischerobryum eurybrochis* (Ren. & Card.) Fleisch., Musci Fl. Buitenzorg 4: 1663. 1923. — Type: Indonesia. Java, Forêt de Tjibodas, sur les pierres de la cascade de Tjibeurreum, 1894–1895 *Massart 1234* (BM-Bescherelle, syntype).

Philonotis etessei Broth. & Paris, Öfvers. Förh. Finska Vetensk.-Soc. 51 A(17): 20. 1909 ("*Etessei*"), *syn. nov.* — New Caledonia. Tipendje, ad rupes, 1902 *Etesse* (H-BR, lectotype, *nov.*).

P. setosa Broth. & Par., Öfvers. Förh. Finska Vetensk.-Soc. 52 A(11): 22. 1911, *syn. nov.* — Type: New Caledonia. Dent de St. Vincent, 1 425 m, VII.1909 *Le Rat* (H-BR, lectotype, *nov.*).

P. novoguineensis Reim., Hedwigia 69: 126. 1929, *syn. nov.* — Type: Indonesia. Nova Guinea neerlandica: in reg. flum. Mamberamo, in monte Doorman. 3 260 m, 24.X.1920 *Lam 1711* (not seen).

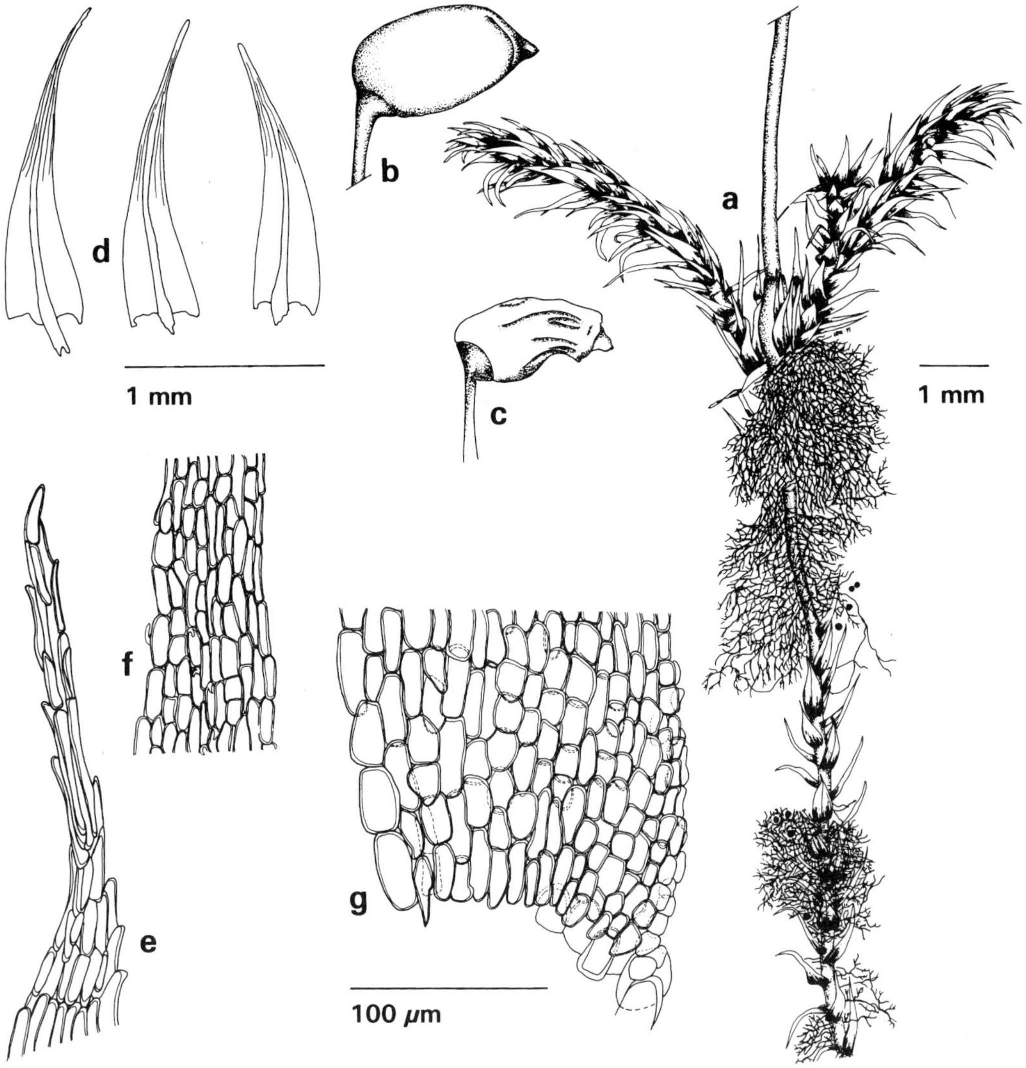

Fig. 7. *Philonotis vescoana* (Besch.) Par. (from *Koponen 32698*, H). — a: Growth habit. — b: Wet capsule. — c: Dry capsule. — d: leaves. — e: Leaf apex. — f: Distal leaf cells. — g: Leaf base. — Use the 100 µm scale for the cellular details, the 1 mm scale on the left for the leaves, and the 1 mm scale on the right for the growth habit.

P. viridifolia Bartr., Occasional Papers Bernice P. Bishop Mus. 11(20): 11, *fig. 4.* 1936, *syn. nov.* — Type: Fiji. Vanua Levu, Mbua, upper Ndama River valley, elevation 100–300 m, on rocks in streams, *Smith* 1695 (FH-Bartram, holotype).

Philonotis hamata Bartr., Lloydia 5: 266, *pl. 2, fig. 23.* 1942, *syn. nov.* —Type: Indonesia. Irian Jaya, Jayawijaya, 9 km NE of Lake Habbema, 2 800 m camp. Abundant on clay of landslip in forest, alt. 2 800 m, X.1938 *Brass 10604* (FH-Bartram, holotype).

Plants 1–3 cm tall, pale, in loose tufts; stems kept together by tomentum, branched with subfloral innovations, branches stiff; epidermis in stem cross-section with thin outer cell walls (hyalodermis), one or two layers of substereidal cortical cells and thin-walled medullary and central cells; leaves distant, erect to erect spreading, falcate to flexuose when dry, spreading to recurved when moist, smooth, falcate, 0.4–0.6 × 1.3–2.0 mm, acuminate from ovate decurrent base, broadest well above insertion, apex stiff or piliferous; border recurved to revolute at base, plane above, basal marginal cells entire, roundish, border at mid-leaf serrulate by protruding cell cor-

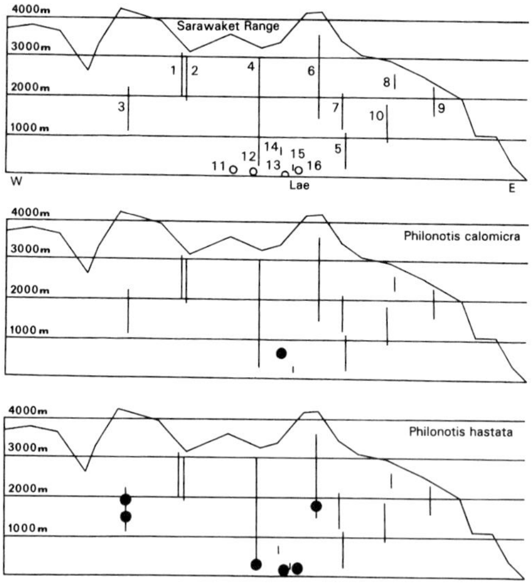

Fig. 8. Altitudinal ranges of *Philonotis calomicra* Broth. and *P. hastata* (Duby) Wijk & Marg. on the Huon Peninsula. Compare with fig. 4 in Koponen and Norris (1983). Nos. 1–16 indicate collecting localities and altitudinal range of the Koponen and Norris expedition.

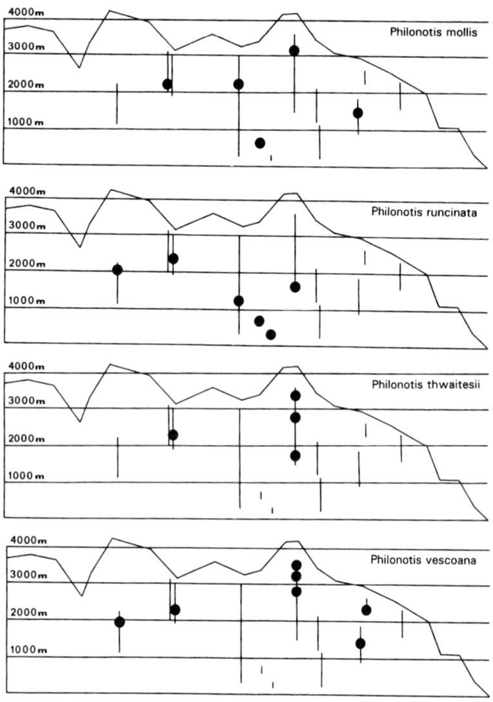

Fig. 9. Altitudinal ranges of *Philonotis mollis* (Dozy & Molk.) Mitt., *P. runcinata* C. Müll. ex Aongstr., *P. thwaitesii* Mitt., and *P. vescoana* (Besch.) Par. on the Huon Peninsula. Compare with Fig. 8 and its legend.

ners, geminate, apical border serrate by larger single or double teeth; costa brownish, 50–80 µm at base, smooth on ventral side, excurrent into the stiff serrate or piliferous point, epidermis cross-section smooth to mammillose, a few conductive cells and dorsal stereidal cells; cells at mid-leaf rectangular to elongate, 10–15 × 15–30 µm, thin-walled or with slightly thickened walls, broadly mammillose apical cells slightly narrower than mid-leaf cells, median basal cells wider and longer, thin-walled, 10–25 × 22–87 µm, rectangular to elongate rhomboidal.

Dioicous. Perichaetia terminal, inner leaves from ovate and concave base abruptly contracting to filiform apices; perigonia gemmiform, leaves 1.6–2.0 mm long, base cochleariform, abrubtly contracted to acute-acuminate apex; seta ca. 3 cm long, capsule horizontal and sulcate whole of its length when dry, gibbous when wet, 1.6–2 mm wide and 2.0–3.5 mm long, constricted below mouth when dry; operculum conic with a short point; exothecial cells 25–67 µm, thick-walled, quadrate, hexagonal or elongate; peristome teeth yellowish, 0.6 mm long, papillose; endostome pale, segments with ciliate apices, as long as exostome or longer, basal membrane about half of endostome length, cilia not seen. Spores ca. 25 µm, papillose in light microscope.

Illustrations: Bartram 1936: 11 (fig. 4, as *Philonotis viridifolia*); Bartram 1942: 265 (fig. 23, as *P. hamata*); Whittier 1976: 194 (fig. 53).

Philonotis vescoana differs from other New Guinea species of *Philonotis* in many characters. Its leaves are distantly arranged so that the reddish brown stem is visible between the falcate and decurrent leaves, which when wet are spreading or even recurved. The lax areolation and low and broad mammillose basal leaf cells are good diagnostic characters, as well as the comparatively large teeth at the apical margin of the leaves. The widest part of the leaf is above the insertion and the costa is excurrent in a stiff or piliferous serrate point, which separates *P. vescoana* from *P. hastata* which has similarly wide and low mammillose leaf cells, but has greenish, much narrower and weaker costa which is rarely excurrent.

Philonotis eurybrochis (= *P. vescoana*) was included in the genus *Fleischerobryum* by Fleischer (1923) and Kabiersch (1938) on the basis of its narrow costa and thin-walled parenchymatous laminal cells. The fertile gametophyte and sporophyte were not known, and the only known specimens came from two sites from the island of Java. Many of our specimens, and the type of *P. hamata*, contain sporophytes. The horizontal and gibbous capsule is closer to the capsules of *Philonotis* than to the genus *Fleischerobryum* which differs in having a horizontal elongate-ovate capsule with a distinct tapering apophysis.

Fleischer (1904) discussed the variation of *Philonotis eurybrochis* (= *P. vescoana*) and concluded that the variation is due to habitat. On dry sites the plants are small, they have narrower leaves and the leaf cells are small, thick-walled and mammillate. This variation also occurs in our material, e.g. in the type of *P. hamata* Bartr. The gametophytes are rather small and the cells narrower than in most other specimens. The habit of the plants, falcate distant leaves and the costa are similar to some other *P. vescoana* specimens, and the capsule being sulcate for the whole of its length also fits. We think that the narrower cells may be due to dryer habitat: "on clay of landslip". Most other specimens were taken on wet localities (see below). The second record of *P. hamata* from New Guinea by Bartram (1959, *Robbins 751*, FH) is a typical plant of *P. vescoana*. Two of the specimens from Fiji (*Norris 44695*, *44501*) are similar to the type of *P. hamata*, and the third (*Norris 44377*) is rather typical *P. vescoana*.

The synonymization of *Philonotis novoguineensis* Reim. is based on the protologue which fits well except that Reimers (1930) described the leaf base as not decurrent, but our plants have decurrent leaf bases. The only other record of *P. novoguineensis* was tentatively given by Bartram (1965; *Brass 30407*, FH) and we confirm his identification.

On the Huon Peninsula *Philonotis vescoana* grows on wet soil or peat at lake sides and on boulders along streams, and was once collected on soil beside trail, and once in a sinkhole on moist soil, at the altitudes between 1 450–3 550 m. The Streimann specimens were taken from similar habitats, the lowest altitude being 1 300 m.

Range on the Huon Peninsula: *2a*. 33629. *2b*. 33426. *3l*. 61094. *6a*. 31557, 31561, 31648, 62207, 62210. *6b*. 62663. *6e*. 32222. *6j*. 63288. *6k*. 32683, 32698, 32704. *6u*. 62749. *8g*. 31410. *10a*. 28312.

Range in Western Melanesia (first reports): West Irian. *4*, *7* (types of *Philonotis novoguineensis* and *P. hamata*). *12*. Present report, and: Ekuti Divide, Bulolo-Aseki Road 33 km WSW of Bulolo, on semi-exposed road cutting, 2 260 m, *Streimann 20261* (H); Yakwoi R., 2 km NW of Menyamya, regrowth beside rocky river, on shaded ground beside stream, 1 300 m, *Streimann 19248* (H). *14*. Mur Mur Pass, 7 km NE of Tambul. Disturbed *Cryptocarya-Elaeocarpus* dominated forest on ridge, on swampy exposed ground, 2 800 m, *Streimann 21242, 21243* (H). *15*. Junction Keglsugl and Bundi Roads, 9 km SE of Mt Wilhelm, grasslands with small regrowth shrubs and scattered *Nothofagus*, on side of fast flowing stream, 2 350 m, *Streimann 18549, 18550* (H); Mt. Wilhelm, above Lake Aunde, 12 000 ft., 1968 *Weber & McVean* B32003 (NY). *17*. Tari Gap, 29 km ESE of Tari, on Tari–Mendi Road, grasslands in a valley, on stream bank, 2 560 m, *Streimann 32771* (H).

Other specimens studied. — Fiji (new country record). Viti Levu, Ba Prov., near Nadarivatu, along road in dense second-growth rainforest, on moist soil, 800 m, *Norris 44377*, Mt. Lomalagi E of Nadarivatu, on moist soil on trail in dense rainforest, 800–900 m, *Norris 44501*, Mt. Victoria, dense rainforest, on moist shaded boulder, 1 000–1 200 m, *Norris 44695* (H). Society Islands. Tahiti, District of Papenoo, 1909 *Tilden 71B* (E, "*Philonotis tildeniana* Cardot sp. nova."). Vanuatu (new country record). New Hebrides Mosses, on Santo, leg. *Rev. F. G. Bowie* (H-BR, herb. Watts no. *49*).

Total range: As 4: Ind (Java, West Irian) PNG. Oc: Fij NC Soc Van.

Acknowledgements. We wish to thank H. Streimann (CBG), W. A. Weber (COLO) and the Curator in LAE for use of their specimens collected in New Guinea. We also are grateful to W. Buck (NY), A. Eddy (BM), P. Isoviita (H), D. Long (E), D. Pfister (FH), and A. Touw (L) for loan of critical type material, and to Mrs. Marjo Urbański and Outi Vainio for drawing the figures. The grant to Senior Scientists in 1989–90 and 1995–96 from the Academy of Finland to the author Koponen for this project is cordially acknowledged.

REFERENCES

Akiyama, H. & Nishimura, N. 1993: The Climacium-type branch development in the Bartramiaceae and its taxonomic significance. — Bryologist 96: 185–191.

Bartram, E. B. 1931: Mosses of Raiatea. — Occas. Pap. Bernice P. Bishop Mus. 9(16): 1–14.

Bartram. E. B. 1933: Manual of Hawaiian mosses. — Bishop Mus. Bull. 101: 1–275.

Bartram, E. B. 1939: Mosses of the Philippines. — Philippine J. Sci. 68: 1–437, pls. 1–29.

Bartram, E. B. 1959: Contribution to the mosses of the Highlands of eastern New Guinea. — Brittonia 11: 86–98.

Bartram, E. B. 1961: Mosses of the Western Highlands of Eastern New Guinea. — Rev. Bryol. Lichénol. 30: 185–207.

Bartram, E. B. 1965: Mosses of the Eastern Highlands, New Guinea, from the 6th Archbold Expedition, 1959. — Contr. U. S. Natl. Herb. 37: 43–67.

Bosch, R. B. van den & Sande Lacoste, C. M. van der 1855–

1870: Bryologia Javanica seu descriptio muscorum frondosorum archipelagi Indici iconibus illustrata 1–2. 1 (1855–1861): I–IV, 1–161, tab. I–CXXX; 2 (1861–1870): I–IV, 1–238, tab. CXXXI–CCCXX. — E. J. Brill, Ludguni-Batavorum.
Brotherus, V. F. 1924: Musci (Laubmoose). II. Spezieller Teil. — In: Engler, A. & Prantl, K. (eds.), Die natürlichen Pflanzenfamilien, ed. 2., 10 (1): 129–131, 143–478. W. Engelmann, Leipzig. 478 pp.
Brotherus, V. F. 1926: Contributions to the bryological flora of the Philippines, VI. — Philippine J. Sci. 31: 277–300.
Chopra, R. S. 1975: Taxonomy of Indian mosses (An introduction). — Bot. Monogr. 10: I–XL, 1–631. CSIR, New Delhi.
Dixon, H. N. 1942: Papuan mosses. — J. Bot. 80: 1–35.
Dixon, H. N. 1943: Alpine mosses from New Guinea. — Farlowia 1: 25–40.
Dozy, F. & Molkenboer, J. H. 1854: Musci frondosi. — In: Miquel, F. A. W. (ed.), Plantae Junghuhnianae. Enumeratio plantarum, quas in insulis Java et Sumatra detexit Fr. Junghuhn 3: 312–341.
Dozy, F. & Molkenboer, J. M. (fortgesetzt Von R. B. van den Bosch & C. M. van der Sande Lacoste) 1855–70: Bryologia Javanica seu descriptio muscorum frondosum Archipelagi Indici. 1–2. — E. J. Brill, Leiden. 1: 161 pp., pls. 1–130; 2: 238 pp., pls. 131–320.
Fleischer, M. 1904: Die Musci der Flora von Buitenzorg (zugleich Laubmoosflora von Java). 2: I–XVIII, 381–643. E. J. Brill, Leiden.
Fleischer, M. 1923 ("1915–22"): Die Musci der Flora von Buitenzorg (zugleich Laubmoosflora von Java) 4: I–XXXI, 1105–1729. E. J. Brill, Leiden.
Frahm, J.-P. 1976: Philonotis hastata (Dub.) Wijk et Marg. neü fur Europa. — J. Bryol. 9: 17–19.
Gangulee, H. C. 1974: Mosses of eastern India and adjacent regions 4: XX–XXIV, XVI, 831–1134. — Publ. by the author, Calcutta.
Griffin III, D. & Buck, W. R. 1989: Taxonomic and phylogenetic studies on the Bartramiaceae. — Bryologist 92: 368–380.
Hegewald, E. & Zanten, B. O. van 1986: A list of bryophytes from Bali (Indonesia) collected by E. and P. Hegewald in 1981. — J. Hattori Bot. Lab. 60: 263–269.
Iwatsuki, Z. 1977: Notes on Philonotis hastata (Duby) Wijk et Marg. in Japan. — Proc. Bryol. Soc. Japan 2(2): 13–15.
Iwatsuki, Z. & Mizutani, M. 1972: Coloured illustrations of bryophytes of Japan. — Hoikusha Publ. Co. Ltd., Osaka. VIII + 405 pp.
Iwatsuki, Z. & Noguchi, A. 1975: Mosses of Borneo, II. — J. Hattori Bot. Lab. 39: 315–333.
Kabiersch, W. 1937: Studien über die ostasiatischen Arten einiger Laubmoosfamilien. II. — Hedwigia 77: 74–136.
Koponen, T. 1990: Bryophyte flora of Western Melanesia. — Trop. Bryol. 2: 149–160.
Koponen, T. 1993a: Genus Philonotis (Musci, Bartramiaceae) in the Pacific. — XV International Bot. Congress, Abstracts: 222.
Koponen, T. 1993b: The Project "Bryophyte flora of the Huon Peninsula, Papua New Guinea". — Bryol. Times 76: 6–8.
Koponen, T. & Norris, D. H. 1983: Bryophyte flora of the Huon Peninsula, Papua New Guinea. I. Study area and its bryological exploration. — Ann. Bot. Fennici 20: 15–29.
Koponen, T., Norris, D. H. & Piippo, S. 1992: Bryophyte flora of Western Melanesia. A status report. — Bryobrothera 1: 157–160.
Koponen, T., Piippo, S. & Norris, D. H. 1991: Bryophyte flora of the Huon Peninsula, Papua New Guinea. XLVII. Generic and family indexes to parts I–XLVI. — Acta Bot. Fennica 143: 107–111.
Loeske, L. 1910: Studien zur vergleichenden Morphologie und phylogenetischen Systematik der Laubmoose. — M. Lande, Berlin. 224 pp.
Magill, R. E. 1987: Flora of Southern Africa. Bryophyta. 1. Mosses, Fasc. 2. Gigaspermaceae–Bartramiaceae: I–IX, 293–443. — Bot. Res. Inst., Pretoria.
Mitten, W. 1859: Musci Indiae orientalis. An enumeration of the mosses of the East Indies. — J. Proc. Linn. Soc. Suppl. Bot. 1: 1–171.
Newton, M. E. 1973: A taxonomic assessment of Bartramia, Breutelia and Exodokidium on South Georgia. — British Antarctic Survey Bull. 32: 1–14.
Newton, M. E. 1974: A synoptic flora of South Georgian mosses: IV. Bartramia and Breutelia. — British Antarctic Surv. Bull. 38: 59–71.
Noguchi, A. (supplemented by Z. Iwatsuki) 1989: Illustrated moss flora of Japan. 3: 493–742, i–vi. — Daigaku Printing Co. Ltd. Hiroshima
Norris, D. H. & Koponen, T. 1985a: Bryophyte flora of the Huon Peninsula, Papua New Guinea. VII. Trachypodaceae, Thuidiaceae and Meteoriaceae (Musci). — Acta Bot. Fennica 131: 1–51.
Norris, D. H. & Koponen, T. 1985b: Bryophytes from Frieda River, East and West Sepik Provinces, Papua New Guinea. I. Musci. — Ann. Bot. Fennici 22: 367–396.
Norris, D. H. & Koponen, T. 1990: Bryophyte flora of the Huon Peninsula, Papua New Guinea. XXXIII. Leskeaceae and Fabroniaceae (Musci) plus corrigenda and addenda to previous papers. — Ann. Bot. Fennici 27: 1–12.
Ochi, H. 1962, 1963: Contributions to the mosses of Bartramiaceae in Japan and the adjacent regions 1, 2. — Nova Hedwigia 4: 87–108, pls. 38–49; 5: 91–115, pls. 9–26.
Ochi, H. 1970: Notes on moss flora (VII). — J. Japanese Bot. 45: 21–28.
Pócs, T., Mizutani, M. & Piippo, S. 1994: Bryophyte flora of the Huon Peninsula, Papua New Guinea. LXV. Preliminary contributions on Lejeuneaceae (Hepaticae). 1. — Ann. Bot. Fennici 31: 179–190.
Pócs, T., Piippo, S. & Mizutani, M. 1995: Bryophyte flora of the Huon Peninsula, Papua New Guinea. LVI. Preliminary contributions on Lejeuneaceae (Hepaticae). 2. — Ann. Bot. Fennici 32: 259–268.
Redfearn, P. L., Tan, B. C. & He, S. 1996: A newly updated and annotated checklist of Chinese mosses. — J. Hattori Bot. Lab. 79: 163–357.
Reimers, H. 1930: Beiträge zur Bryophytenflora Neuguineas. — Hedwigia 69: 114–136.
Sakurai, K. 1941: Beobachtungen über japanische Moosflora XXIII. Bartramidula, Philonotis und Fleischerobryum in

Nippon. — Bot. Mag. Tokyo 55: 1–16.
Schultze-Motel W. 1974: Die Moose der Samoa-Inseln. — Willdenovia 7: 333–408.
Scott, G. A. M., Stone, I. G. & Rosser, C. 1976: The mosses of Southern Australia. — Acad. Press Inc., London. XV + 495 pp.
Smith, A. J. E. 1978: The moss flora of Britain and Ireland. — Cambridge Univ. Press, Cambridge etc. 706 pp.
Tixier, P. 1986: Bryophyta exotica, VIII. Bryophytes de Nouvelle-Calédonie. — Cryptog. Bryol. Lichénol. 7: 225–234.
Vitt, D. H., Koponen, T. & Norris, D. H. 1994: Bryophyte flora of the Huon Peninsula, Papua New Guinea. LVI. Orthotrichaceae, Macromitrioideae (Bryopsida). — Acta Bot. Fennica 154: 1–94.
Whittier, H. O. 1976: Mosses of the Society Islands. — Univ. Presses of Florida, Gainesville. X + 410 pp.
Zanten, B. O. van 1964: Mosses of the Star Mountains expedition. — Nova Guinea 16: 263–368.

INDEX TO NAMES

An asterisk means that the lectotype has been proposed in this paper. The accepted names of taxa occurring in Western Melanesia are in **boldface**. The names proposed as new synonyms are marked *syn. nov.*

Acta Bot. Fennica 156: 23–48 ISBN 951-9469-50-8 ISSN 0001-5369
Helsinki 4 September 1996

Bryophyte flora of the Huon Peninsula, Papua New Guinea. LVIII. *Hypnodendron* (Hypnodendraceae, Musci)

Daniel H. Norris & Timo Koponen

Norris, D. H. & Koponen, T., Department of Ecology and Systematics, Division of Systematic Biology, P.O. Box 7, FIN-00014 University of Helsinki, Finland

Received 15 August 1995

In Western Melanesia the family Hypnodendraceae Broth. has 11 species in *Hypnodendron* (C. Müll.) Mitt.: *H. auricomum* Broth. & Geh., *H. brevipes* Broth., *H. dendroides* (Brid.) Touw, *H. diversifolium* Broth. & Geh., *H. fuscomucronatum* (C. Müll.) Jaeg., *H. junghuhnii* (C. Müll.) Jaeg., *H. menziesii* (Hook.) Par., *H. milnei* Mitt., *H. reinwardtii* (Schwaegr.) Jaeg., *H. subspininervium* (C. Müll.) Jaeg. and *H. vitiense* Mitt. The placement of the macronematal apparatus is shown to be an effective diagnostic feature for *Hypnodendron*, section *Phoenicobryum* (Lindb. ex Broth.) Touw. The cross-sectional anatomy of the sporangial wall is shown as a feature distinguishing *Hypnodendron* section *Sciadocladus* (Kindb.) Touw from all other members of the genus. The lectotypes of *H. auricomum*, *H. brevipes*, *H. diversifolium*, and *H. macgregorii* Broth. & Geh. are newly designated. The paper includes keys to the species, descriptions, illustrations and distributional data for the species in Western Melanesia. For each species there are discussions on taxonomy and classification, as well as nomenclatural notes and range descriptions.

Key words: Hypnodendraceae, *Hypnodendron*, macronemata, nomenclature, Papua New Guinea, Solomon Islands, sporangial anatomy, taxonomy, West Irian

CONTENTS

INTRODUCTION

This paper is part of a series dealing with the bryoflora of Western Melanesia (Papua New Guinea, West Irian, and the Solomon Islands). It is based upon the 1981 collection of T. Koponen and D. H. Norris on the Huon Peninsula, Papua New Guinea, and on extensive herbarium study of other Western Melanesian specimens. For the Norris and Koponen collections we list the collection localities in abbreviated form; we have also codified the presentation of the geographic ranges. Explanation of this format is in the first and seventh parts of this series (Koponen & Norris 1983, Norris & Koponen 1985a; see also Norris & Koponen 1990). In addition to our own specimens, we cite here specimens sent from H. Streimann (CBG).

Koponen (1990, 1993), Koponen *et al.* (1992) and Piippo *et al.* (1994) have dealt with the project and its results. The previous parts of the Huon Peninsula series are listed in Koponen *et al.* (1991) and Koponen (1993). The previous paper in the series is Koponen and Norris (1996).

We have not studied all the type specimens cited in the text, but trust Touw's (1971) synonymization of the names. However, the nomenclatural references have been checked against the original literature.

The voucher specimens of our New Guinea studies are in H (1st set) and Lae (2nd set). In addition, specimens of *Hypnodendron* were distributed as exchange to about 70 herbaria, including all major bryological repositories.

FAMILY HYPNODENDRACEAE BROTH. 1909

This family was placed by Brotherus (1909, 1924, 1925) in the Eubryales in affinity with such families as the Rhizogoniaceae, Bartramiaceae, Meesiaceae and Spiridentaceae. Brotherus assigned four genera to this family. Touw (1971) reduced two of these (*Sciadocladus* Kindb., and *Mniodendron* Kindb.) into synonymy under *Hypnodendron*. Touw questioned the placement of *Braithwaitea* Lindb. in the Hypnodendraceae, and Buck and Vitt (1986) transferred that genus into their new family Trachylomataceae.

We find ourselves in substantial agreement with the conclusions of Touw (1971), and of Buck and Vitt (1986). The removal of the Hypnodendraceae from the Eubryales separates from this latter, basically acrocarpous complex the only group which possesses pseudoparaphyllia. We are somewhat uncomfortable, however, because the family consistently has macronemata. We know of no other group of mosses which has both macronemata — normally a feature of the Eubryales *sensu* Brotherus (1924, 1925) — and pseudoparaphyllia (a feature of the Hookeriales, Hypnobryales and Isobryales *sensu* Brotherus). It is of especial interest that most members of the genus *Hypnodendron* have macronematal apparati surrounded by a cluster of pseudoparaphyllia.

Genus **Hypnodendron** (C. Müll.) Mitt. 1873

Hypnodendron (C. Müll.) Mitt. in Seeman, Fl. Vitiensis 10: 401. 1873. — *Hypnum* sect. *Hypnodendron* C. Müll., Syn. Musc. Frond. 2: 496. 1851.

Dendro-Hypnum Hampe, Nuovo Giorn. Bot. Ital. 4: 289. 1872.

Note: Hampe's *Dendro-Hypnum* antedates *Hypnodendron* (C. Müll.) Mitt. by one year. To avoid changes of well-established names, *Hypnodendron* will be proposed to be conserved (Koponen, T., Isoviita, P. & Touw, A., in prep.).

This primarily Southern Hemisphere genus extends north of the equator only in SE Asia. It is such a spectacular plant that it has been well-collected throughout its range, and many of those collections have even come from people who do not normally collect bryophytes. For this reason, the work of Touw (1971) is likely to be a conclusive study of the genus without significant likelihood of the finding of new species, and we follow his treatment. His monograph is, in its care and thoroughness, one of the best examples of monographic work present in the bryological literature. However, we refer to Akiyama's (1989) proposal to reevaluate *Hypnodendron opacum* Fleisch. at the specific level. As a difference to Touw's (1971) treatment we propose several lectotypes instead of holotypes in case of divided authority (e.g. for *Hypnodendron auricomum*). In some other cases (e.g. *H. climacioides*) the protologue cited more than one specimen and then there are syntypes in question, not the holotype.

The genus presents serious difficulties to the taxonomist — largely resolved by Touw (1971). It has little variation in the form of the peristome; the intraspecific variation in the form of the leaves seems greater in many respects than much of the inter-specific variation; the variation in the size of an apparently adult plant in some species may be such that the smaller specimens may be only about one-tenth the height of the larger specimens. Touw utilized characters not traditional in most other groups of bryophytes. Despite unexcelled descriptions and diagrams, one must supplement Touw's work with studies of accurately identified examples of many of the species. Only then have we been able fully to understand some of Touw's important characters of diagnosis. Fortunately, the eleven species in the flora of Western Melanesia can be keyed with greater ease than can the twenty five species in Touw's (1971) work.

In addition to the characters routinely included in the work of Touw (1971), we have found value in describing the shape and placement of the macronematal apparatus. We have rather loosely used the term 'macronematal apparatus' despite the presence of pseudoparaphyllia rather than rhizoids in the halo around the branch initials. This feature is best investigated on the upper portion of the stipe, and observation is facilitated by the removal of all leaves from that stem fragment followed by a longitudinal cut of the leafless stem. The macronematal apparatus is then seen relative to the outer cortical cells as a cluster of thinner, less pigmented cells surrounded in *Hypnodendron* by a halo of pseudoparaphyllia. We have routinely described shape of the macronematal apparatus as a length (long axis of the stem)/width (transverse axis of the stem) ratio. The longitudinal placement of the macronemata on the stem may be 'pseudoaxillary', 'intercalary' or 'abaxial' on the merophyte. The lateral placement of the macronemata may be on the imaginary line decurrent from the costa, on the imaginary line decurrent from the lateral margin of the leaf, or it may be between those lines. An understanding of this positioning requires description of the basic unit of stem anatomy, the 'merophyte'. This latter term refers to the separate products of division from each of the three cutting faces of the tetrahedral apical cell. Each merophyte thus ideally comprises one hundred twenty degrees of stem cross-section although one or more of the merophytes may be enlarged at the expense of the neighboring derivatives. The base (proximal portion) of the merophyte begins at the line of leaf insertion, and it extends to, and is crowned by its individual leaf at its distal end. Thus, a macronematal initial which seems to be placed in the axil of a leaf is truly 'pseudoaxillary', because that initial belongs not to

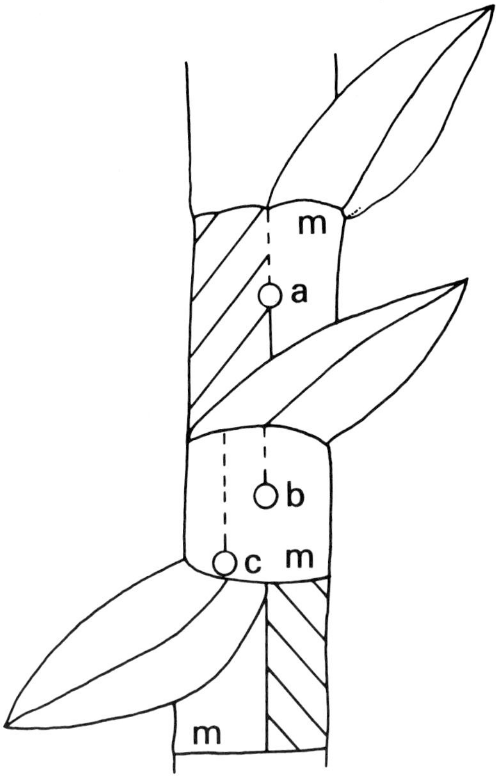

Fig. 1. Schematic presentation of the placement of macronematal apparatus. — a: Lateral. — b: Medial. — c: Pseudoaxillary. m = merophyte.

the leaf which subtends it, but to the leaf distal to it on that same merophyte. Furthermore, we call 'lateral' the placement of macronematal apparatus on the imaginary line drawn downward from the margin of the next more distal leaf, and we call 'medial' those apparatus which are inserted on the imaginary line drawn downward from the costa of the next more distal leaf (Fig. 1).

Touw (1971) described a large number of sections within *Hypnodendron*. Several of these sections should be commented upon. The species of the section *Phoenicobryum* (*Hypnodendron subspininervium*, *H. brevipes*, and *H. reinwardtii*) may be so pinnately branched and decumbent as to obscure the dendroid form which we associate with the genus. In addition to the three species named above, *H. vitiense* may also have such an almost non-dendroid form, and especially poorly developed examples of most of the other species may approach such a pattern. Among the features most definitive of the section *Phoenicobryum* is an apparently axillary placement of the macronemata. Such macronemata are median relative to the leaf which subtends them, and they are located so close to the line of leaf insertion as to seem almost to touch that line. Because such macronemata are truly basal on the merophyte, it is necessary also to relate their position to the leaf above. Typically, the macronemata of *Phoenicobryum* are median and proximal to the subtending leaf but, like other members of the genus, it is located close to the imaginary line of decurrency of the next more distal leaf. Thus, the macronemata of *Hypnodendron* located on the lateral margins of the merophyte may be basal (*Phoenicobryum*) or intercalary (other members of the genus).

Phoenicobryum can also be recognised by its keeled and asymmetric leaves which basally are also plicate and describe a rather puckered concavity. Normally the stipe leaves of *Phoenicobryum* cannot be laid flat upon the slide because of this keeling. Microphyllous and caducous branch leaves may be restricted to the *Phoenicobryum* complex.

Sciadocladus Lindb. ex Broth. was one of the more widely accepted of the several genera in the family until Touw (1971) reduced it to sectional status. We agree with the treatment by Touw, but we note that the traditional differentiating characters (smooth capsule and conic operculum) can be supplemented by a description of the capsule wall cross-section. *Sciadocladus* has a wall-thickening of only the single outermost layer of exothecial cells. Other members of *Hypnodendron* in New Guinea have 2–6 layers of thick-walled cells in a capsule cross-section. The exothecial cells of *Sciadocladus* in surface view may be as wide as 45 µm in comparison with the 20 µm maximum size shown in other members of *Hypnodendron*. Axillary hairs of *Sciadocladus* are not offset from the leaf insertion and they have only one basal brown cell and a single terminal cell; those of the other *Hypnodendron* are minutely offset and they consistently have two basal brown cells and one or two enlarged terminal ones. Our description of the axillary hairs of *Sciadocladus* was made especially difficult by the extremely deciduous character of those structures. In contrast, the hairs of the other *Hypnodendron* are very persistent.

Key to the species of *Hypnodendron* in Western Melanesia

In the following key, we have attempted to use gametophytic characters except when sporophytic characters are absolutely necessary. Because older examples of many of the species often have the stipe denuded, we are emphasizing characters of the branch leaves in at least the initial steps of keying. While we routinely do not recognise inra-specific categories, our treatment of some of the species is limited to those expressions (subspecies and varieties) present in New Guinea.

1. Stipe leaves completely appressed to stem, or with slightly divergent apices 2
— Stipe leaves squarrose or, at least, spreading from the stem so that the majority of the leaf is distant from that stem 5
2. Stipe usually densely covered by rhizoids nearly to the base of the umbellate crown; alar cells of branch leaves abruptly expanded, more than twice as broad as the adjacent laminal cells, arranged in a rather dark green group extending nearly to the costa; branch leaves never with geminate teeth .... 5. *H. fuscomucronatum*
— Stipe mostly with very sparse rhizoids except at its base; alar cells of branch leaves not or little differentiated, not in a well-defined dark pocket of enlarged cells; branch leaves mostly with at least a few geminate teeth 3
3. Stipe so covered with closely appressed, often bleached leaves as to obscure nearly all of its surface; plant mostly with umbellate branching .... 1. *H. auricomum*
— Stipe with remote, never strongly bleached leaves; plant mostly with pinnate branching 4
4. Branch leaves with very sharp teeth mostly over 50 µm long; geminate teeth scarce on both branch and stem leaves; stipe leaves dentate to near the base .... 6. *H. junghuhnii*
— Branch leaves with teeth blunt and nearly always geminate, seldom more than 25 µm long; stipe leaves entire below the middle .... 11. *H. vitiense*
5. Branch leaves with marginal teeth singly placed 6
— Branch leaves with marginal teeth mostly geminate .. 9
6. Stipe leaves so densely covered with rhizoids as to hide at least the basal one-half of the leaves; margin of stipe leaves finely dentate on the acumen but nearly entire below except for a few rather coarse dentations on the well-defined auricle .... 3. *H. dendroides*
— Stipe free of rhizoids except at its base, or so loosely clothed with rhizoids as to allow visibility even of the leaf base; margin of stipe leaves coarsely to finely dentate above but without coarse teeth at the leaf insertion 7
7. Branch leaves complanate and tristichous with the laterally inserted leaves relatively broader and somewhat longer than the more medially inserted leaves; base of stipe leaves at least somewhat decurrent, often prominently so .... 4. *H. diversifolium*
— Branch leaves isomorphic, or with more distal leaves gradually smaller than the more proximal ones; base of stipe leaves never long-decurrent 8
8. Alar cells of branch leaves incrassate and porose, yellow to brown; margin of branch leaves with coarse multicellular teeth distantly placed and alternating with low serrulation; branch leaves ovate-lanceolate, 2–2.5: 17. .... 7. *H. menziesii*

— Alar cells of branch leaves not strongly differented, concolorous or somewhat more pale than adjacent laminal cells; margin of branch leaves with regularly-placed teeth formed of only a distal portion of a single cell; branch leaves lanceolate, 3–5: 1 10

9. Plants mostly palmate above a tall stipe; alar cells of branch leaves abruptly differentiated in a sometimes excavate pocket; enlarged and porose; margins of branch leaves with very sharply pointed teeth, mostly geminate, mostly more than 50 µm long .... 8. *H. milnei*

— Plants mostly pinnate with stipe poorly defined; alar cells poorly differentiated, often yellowish, little enlarged relative to adjacent cells 10

10. Leaves of the larger branches ovate, about 2.5: 1; well-developed plants bipinnate; urn inclined to horizontal .... 10. *H. subspininervium*

— Branch leaves lanceolate to elliptic lanceolate; 3–5: 1; urn mostly erect 11

11. Seta up to 3 cm long; urn up to 0.5 cm long, shallowly sulcate .... 2. *H. brevipes*

— Seta and urn longer; urn often deeply sulcate ...... 9. *H. reinwardtii*

1. *Hypnodendron auricomum* Broth. & Geh. (Figs. 2, 12)

Öfv. Finska Vet. Soc. Förh. 40: 190. 1898. — Type: Papua New Guinea. Milne Bay, Mt. Dayman, *W. Armit junior 622* (H-BR!, lectotype, *nov.*).

Sciadocladus novae-guineae Dix., Farlowia 1: 34. 1943. — *Hypnodendron novae-guineae* (Dix.) Bartr., Brittonia 9: 41. 1957. — Type: Papua New Guinea. Central, Alola, 6 000 ft., *Carr 14184* (BM, holotype, not seen). — Synonymized by Touw (1971).

Sciadocladus wisselii Dix., Farlowia 1: 34. 1943. — Type: Indonesia. West Irian, Jayawijaya, Mt. Carstensz, 3 800–4 300 m, *Wissel 138* (BM, not seen). — Synonymized by Touw (1971).

Hypnodendron climacioides Nog., J. Hattori Bot. Lab. 10: 12, *fig. 6*. 1953. — Types: Papua New Guinea. Morobe, Mt. Sarawaket, 3 350–3 560 m, V.1939 and 15.IV.1939 *Clemens* (TNS, syntypes, not seen). — Synonymized by Touw (1971).

Plants to 12 cm high, mostly palmate to pinnate but occasionally umbellate, with branches to 5 cm long, golden-green to yellow green, often glossy, usually erect, with a well-defined stipe which has imbricated, mostly strongly appressed leaves with few rhizoids except at its base. Upper stipe leaves often bleached white, symmetric, to 4 mm long, mostly 1.6–2:1 but sometimes to 3:1, keeled only near apex, distant, mostly ovate but occasionally ovate-lanceolate to lanceolate, broadest immediately above their bases, mostly blunt at apex, but occasionally acute to acuminate. Ultimate branch leaves approximate to imbricated, symmetric, usually somewhat complanate, sometimes smooth but mostly lightly rugose when dry, erect to ascending, sometimes slightly secund, to 2 mm long, mostly about 2: 1, broadest about 1/4 from base and gradually contracted to a very short acumen. Median laminal cells to 6 µm broad, mostly more than 10:1, mostly lightly to very strongly prorate at both ends but occasionally essentially smooth, mostly rather thick-walled, with lumen/wall ratio 1–2:1. Alar cells not differentiated but basal cells distinctly larger and thicker-walled than median laminal cells, usually orange to red-brown. Costa with a few rather large dorsal teeth, occasionally percurrent but mostly ending within a few cells of leaf apex. Marginal cells shorter than interior cells with many of them only 1–2:1. Margins plane throughout, dentate in distal 3/4, mostly with large multicellular teeth alternating with irregularly arranged smaller teeth; geminate teeth infrequent to absent. The larger teeth, especially those of stipe leaves, often deciduous. Leaves decurrent with decurrency often extending to next leaf. Axillary hairs not offset from leaf insertion, to 70 µm long, with 2 basal brown cells and 12 terminal cells. Pseudoparaphyllia lanceolate and erose. Macronematal apparatus scattered but frequent on the outside of the decurrency of the stipe leaves, elliptical, to 50 µm wide, 5–6:1, to 3 cells wide and 6 cells long, with rhizoid initials mostly clustered on the acroscopic end above the clear cells of the bud, and with pseudoparaphyllia often associated with that bud. Stem rounded pentagonal in cross section, with a prominent central strand and with the largest inner cortical cells isodiametric and rather thin walled, to 25 µm, grading imperceptibly to several rows of thick-walled, reddish outer corticals, to 8 µm in diameter.

Dioicous. Seta to 5 cm long but often as short as 2 cm, red-brown, mostly at least somewhat flexed distally. Seta cross-section rounded quadrangular with three layers of brown pachydermous cells well differentiated from the white, moderately pachydermous inner cells and from the strong central strand. Urn to 6 mm, cylindric, about 6:1, mostly neither arcuate nor sulcate. Operculum rostrate, about 1/2 as long as urn. Capsule wall to 120 µm thick with up to 5 layers of cells. Exothecial cells rectangular, 3–8:1, not in organized streaks of broad and narrow cells but with scattered broad (to 20 µm) cells

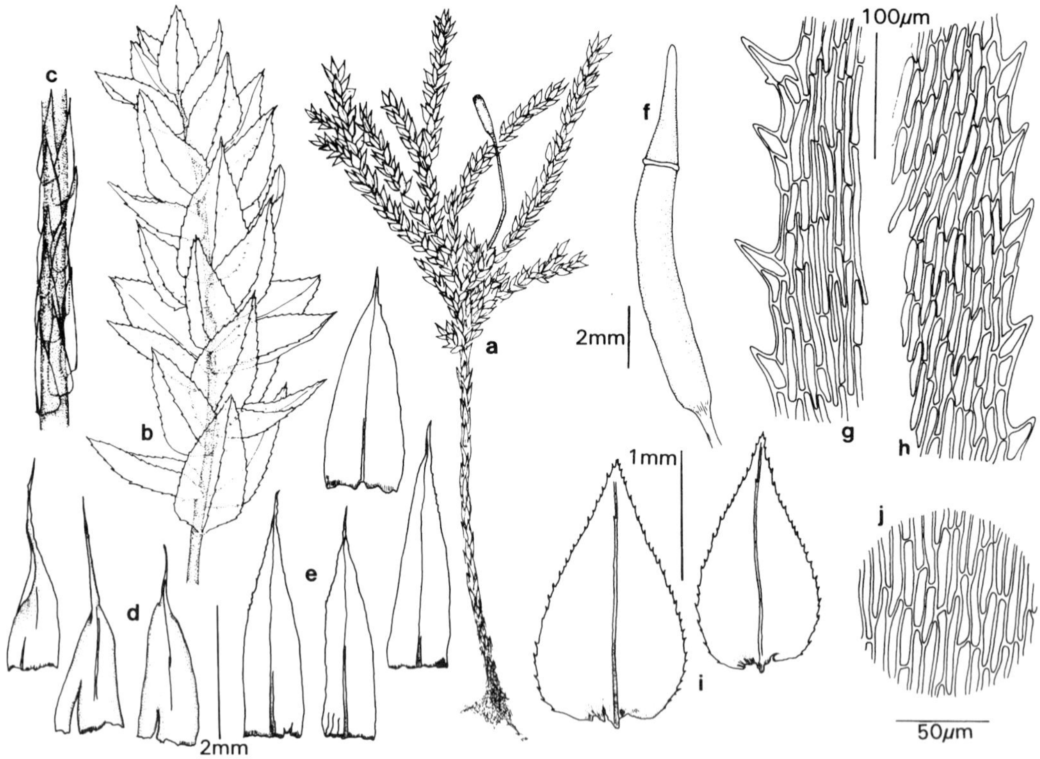

Fig. 2. *Hypnodendron auricomum* Broth. & Geh. (a–c from *Norris 64143*, UC; d–j from *Norris 65488*, H). — a: Growth habit. — b: Branch. — c: Stipe. — d: Perichaetial leaves. — e: Stipe leaves. — f: Capsule. — g: Leaf border of stipe leaf. — h: Leaf border of branch leaf. — i: Branch leaves. — j: Leaf cells. — Use the 2 mm scale on the left for d and e, the 2 mm scale on the right for f, the 1 mm scale for i, the 100 µm scale for g and h and the 50 µm scale for j.

in a matrix of narrower (to 10 µm cells); little differentiated at mouth into isodiametric cells. Stomata superficial, restricted to short neck of capsule. Annulus vesicular, mostly not revoluble. Exostome to 800 µm long, dorsal plate more narrow than ventral one, horizontally closely striate, with a zig-zag median commisural line. Endostome sparsely papillose, nearly as long as the exostome, with basal membrane comprising 1/2–2/3 of its length, processes keeled and perforate, cilia 2–5. Spores smooth or nearly so, to 25 µm.

Illustrations: Noguchi 1953: 13 (fig. 6) as *Hypnodendron climacioides*; Touw 1971: 254 (fig. 12), 256 (fig. 13)

This is a puzzlingly variable species which includes some of the largest specimens assignable to the genus, and it also includes some very small specimens. Touw (1971) points out that these small specimens often have very fragile branch leaves, and have papillae so strong and confluent as to give a verrucose macroscopic appearance. We have many plants conforming to this description, and note that the papillar salients are often lobed and they may reach 15 µm in height. The lobing of those papillae seems to appear on the end walls of laterally adjacent cells and they may even be tri to tetra-lobed when several cells end together. Some of these small specimens have sporophytes, and we have noted many differences relative to the larger plants of *Hypnodendron auricomum*. The setae are markedly more slender; and the capsule wall is thinner, being only 100 µm thick and composed of only 3 layers of cells. The spores are smooth to warty papillose, bimodal in size with some green and multicellular, to 50 µm in diameter, and with other not green ones, to 20 µm. Most of the cells of the exothecium are about 12–15 µm and only 2–5: 1. Akiyama (1989) accepted these small forms as a separate species, *H. opacum* Fleisch., and cited two of our collections (*Norris 65488, 66214*). We follow Touw's (1971) treatment, and include *H. opacum* in the circumsciption of *H. auricomum*.

Hypnodendron auricomum can be confused only with *H. fuscomucronatum*. The key in Touw (1971) suggests that the stipe of the former species is "tomentose at base only, completely covered by the usually coarsely serrate leaves"; while the latter is tomentose throughout and is "not completely covered by the leaves". We have many specimens which seem intermediate in both these respects, and we find the large, usually multicellular, often caducous teeth of *H. auricomum* to be a more diagnostic feature. Specimens with obviously complanate branch leaves can be referred to *H. auricomum*. The costa of *H. fuscomucronatum* is usually longer than that of *H. auricomum*, and specimens with the costa minutely excurrent are dependably referred to the former species. The leaves of *H. auricomum* may be irregularly rugose when dry, while those of *H. fuscomucronatum* may be plicate but not rugose. When sporophytes are present, the smooth and nearly erect capsule of *H. auricomum* allows distinction from the sulcate and more cygneous one of *H. fuscomucronatum*.

Hypnodendron auricomum was found at 750–3 550 m mainly in primeval rainforest (35 localities). It was also found in second growth forest (10), savannahs (4) and scrub (1). The substrates of our 104 specimens are: tree (49), soil (21), log (14), tree fern (11), humus (7), and cliff (1).

Range on the Huon Peninsula: *1a*. 34586, 34597. *1e*. 65056, 65103, 65199, 65259, 65260. *2b*. 33314, 33592, 33597. *2c*. 33540. *2d*. 33704. *2e*. 33845. *2i*. 34141. *2o*. 34320, 34331. *2s*. 63919, 63888. *2t*. 64051. *3a*. 29274. *3b*. 29232, 29488a, 29489. *3e*. 29750. *3f*. 29786, 29880. *3h*. 60133, 60171, 60240, 60257, 60320. *3i*. 60492, 60533, 60539. *3j*. 60687. *3k*. 60861. *4b*. 734. *4g*. 66652, 66659. *4i*. 67043. *5c*. 66082, 66083, 66085, 66086. *6a*. 62265, 62381. *6b*. 31825, 31860, 31959, 62655. *6c*. 32001, 32068, 32086. *6e*. 32241, 62899. *6j*. 63250. *6l*. 32795, 32837, 32848. *6p*. 33066, 33081, 33126. *6q*. 33048. *6w*. 62711A, 62773, 62799, 62808, 62817. *6x*. 62972. *6y*. 63328. *6z*. 63549. *6A*. 63578, 63631, 63661, 63775. *7a*. 65301, 65398, 65399. *7b*. 65484, 65488, 65495. *8a*. 31084. *8b*. 30996. *8d*. 31134. *8e*. 31235, 31244. *9d*. 30282, 61642. *9e*. 30356. *9g*. 30676, 30740. *9k*. 61778, 61790. *9l*. 62062, 62063. *10c*. 28583. *10d*. 28711, 28739, 28756, 59521, 59522. *10n*. 59695. *14e*. 27985, 28016.

Range in Western Melanesia: West Irian. *6, 7*. Papua New Guinea. *11*. Present report. *12*. Present report, (H), *13–17*. *18*. Nebilyer River, 2 760 m, on tree fern, *Streimann 20617* (H). *20–23*. Solomon Islands. *31*.

Total range: As 4: Ind PNG; Oc: Sol.

2. *Hypnodendron brevipes* Broth. (Fig. 3)

Öfv. Finska Vet. Soc. Förh. 40: 189. 1898. — Type: Papua New Guinea. Milne, Cloudy Mts, 1 000–4 000 ft., *Micholitz 130 bis* (H-BR!, lectotype, *nov.*).

Hypnodendron laxum Bartr., Brittonia 9: 41. 1957. — Type: Papua New Guinea. Milne Bay, Mt. Dayman, 1 550 m, *Brass 23078* (FH, holotype, not seen). — Synonymized with reservation by Touw (1971).

Plants to 6 cm high, pinnate, sometimes closely so, with branches to 3 cm long, stramineous to green, often with a reddish cast, very glossy, prostrate to ascending with a well-defined stipe which has strongly reflexed leaves and has rhizoids mostly confined to its base. Upper stipe leaves to 3.5 mm long, 2–3:1, asymmetric and distally so strongly keeled as to resist placement flat on a slide, distant to approximate, ovate-lanceolate, often acuminate near apex. Ultimate branch leaves approximate to imbricated, not complanate, not plicate when dry, loosely divergent, ovate-lanceolate, to 3 mm long, 3–4:1, broadest near the base and only gradually contracted to the acumen. Median laminal cells to 6 μm broad, mostly more than 10:1, usually smooth but occasionally slightly prorate, mostly rather thick-walled, with lumen/wall ratio 1–2:1. Alar cells in a poorly defined group, distinctly larger than the median laminal cells, to 25 μm broad, 1–2:1, not arranged in auricles. Costa excurrent in a short cusp, usually with only a few dorsal teeth, or with those teeth more densely placed on the cusp. Marginal cells not markedly shorter than interior cells. Margins recurved at base but plane above, dentate in distal 3/4, with teeth mostly composed of more than one cell, usually with many geminate teeth, especially on branch leaves. Decurrency present on stem, often extending to next leaf, with the leaf margins sometimes confluent in plants with especially short internodes. Axillary hairs slightly offset from leaf insertion, to 120 μm long, with 2 basal brown cells and 1–2 terminal cells. Pseudoparaphyllia on stipe to 80 μm long, to 6 cells long and 2–4 cells wide at base, often inserted on the macronematal apparatus; those on branches mostly filamentous, 2-celled, to 8 μm. Macronematal apparatus pseudoaxillary, present outside the decurrency on one side of nearly every stipe

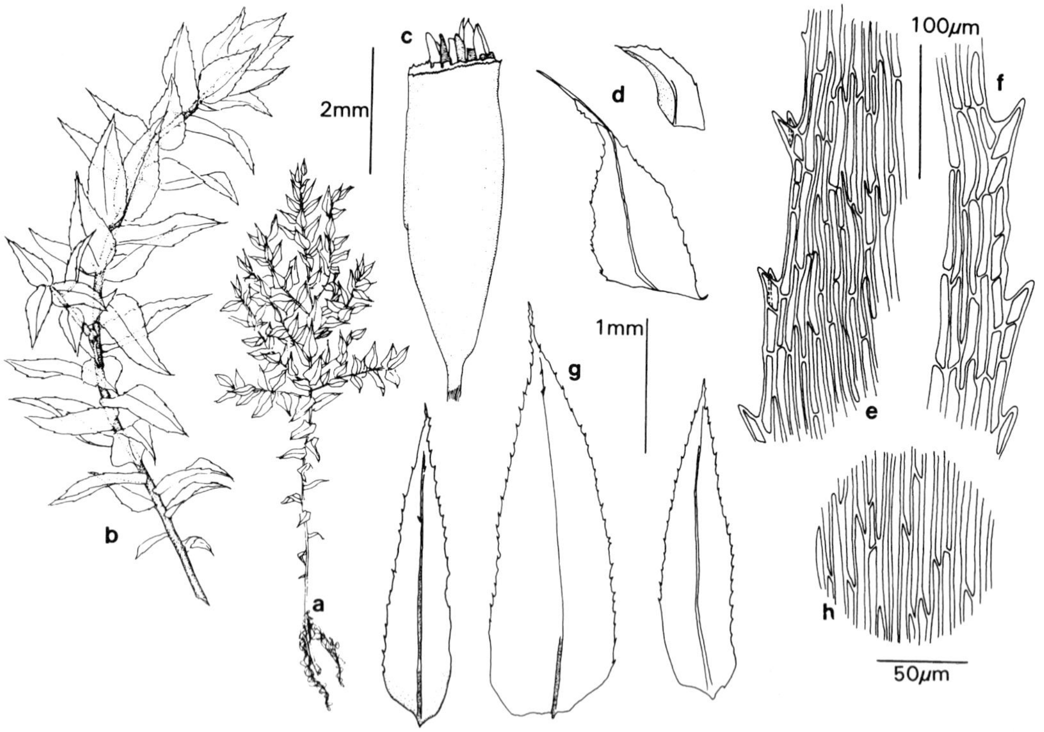

Fig. 3. *Hypnodendron brevipes* Broth. (a, b from *Koponen 27920,* UC; c–h from the lectotype, H). — a: Growth habit. — b: Branch. — c: Capsule. — d: Reflexing of stipe leaves. — e: Leaf border of branch leaf. — f: Leaf border of stipe leaf. — g: Branch leaves. — h: Leaf cells. — Use the 2 mm scale for c, 1 mm scale for d and g, the 100 µm scale for e and f and the 50 µm scale for h.

leaf, elliptical, to 125 µm wide, 2–4:1, to 3 cells wide and 5 cells long.

Dioicous. Seta to 3 cm long, red-brown, mostly at least somewhat flexed distally. Seta cross-section rounded pentagonal with three layers of yellow brown pachydermous cells, 1–2 layers of large, hyaline, leptodermous cells and a moderately well-defined central strand surrounded by small and pachydermous yellow brown cells. Urn to 4.5 mm, cylindric, about 4:1, nearly erect and lightly sulcate to smooth. Operculum rostrate, about 1/2 as long as urn. Capsule wall mostly 3 layers thick. Exothecial cells in vertical rows of broad cells (to 20 µm wide) and narrow cells (to 12 µm wide), rather thin-walled, rectangular, mostly more than 4:1. Stomata superficial, restricted to short neck of capsule. Annulus vesicular, mostly not revoluble. Exostome to 800 µm long, dorsal plate more narrow than ventral one, horizontally closely striate, with a zig-zag median commisural line. Endostome sparsely papillose, nearly as long as the exostome, with basal membrane comprising 1/2–2/3 of its length, processes keeled and perforate, cilia 1–4. Spores smooth or nearly so, to 10 µm.

Illustrations: Brotherus 1909: 1168 (fig. 823 F–H); Brotherus 1924: 436 (fig. 384 F–H); Touw 1971: 248 (fig. 9).

Touw (1971) lists the two syntypes of *Hypnodendron brevipes* but he does not select a lectotype. The two syntypes in H-BR are identical in every gametophytic respect, but we have selected the H-BR specimen of the *Micholitz 130 bis* collection from the Cloudy Mountains. The other specimen (*Micholitz 130*, from mountains near Mita) lacks sporophytes — features included in Brotherus' (1898) original description. Another specimen in H lacks any form of labelling except for the identification, but this appears to have been separated from the lectotype (*Micholitz 130 bis*), and it is annotated: "Drawing made of this specimen published in Brotherus, V.F. 1909, 1924 . . ."

Touw (1971) suggests that *Hypnodendron brevipes* and *H. reinwardtii* are of doubtful distinction. In our key, and that of Touw, the distinctions are all sporophytic: length of seta and urn;

and amount of sulcation. We note, however, that such characters are highly variable in other members of the genus. Certain possible gametophytic differences are worth noting. We have observed that, relative to *H. reinwardtii*, the stipe leaves of *H. brevipes* may be somewhat more abruptly acuminate at mid-leaf, possibly with a longer subula formed by the costa. Another feature of occasional utility is the contortion of the stipe leaves of *H. brevipes* — reminiscent of *H. diversifolium*.

Range on the Huon Peninsula: *Hypnodendron brevipes* was not present among our Huon Peninsula collections.

Range in Western Melanesia: Papua New Guinea. *20. 22. 23.*

Total range: Endemic to Papua New Guinea.

3. *Hypnodendron dendroides (*Brid.) Touw (Figs. 4, 12)

Blumea 19: 320. 1971. — *Bryum dendroides* Brid., Musc. Recent. 2(3): 20. 1803.

Mniodendron divaricatum (Reinw.) Lindb., Öfv. K. Svenska Vet.-Akad. Förh. 18: 375. 1862. — *Hypnum divaricatum* Reinw. in Reinw. & Hornsch., Nova Acta Acad. Leop. Car. 14(2): 723, *tab. 40 d, figs. 1–8.* 1829. — Type: Indonesia. Java, G. Gedeh, *Reinwardt s. n.* (L, lectotype, sel. Touw, not seen). — Synonymized by Lindberg (1862) and Touw (1971).

Mniodendron humile Lindb. in Bosch & Sande Lac., Bryol. Jav. 2: 138, *tab. 235.* 1865. — Synonymized by Touw (1971).

Hypnodendron palmaeum Mitt., Proc. Linn. Soc. N. S. Wales 7: 103, 1882. — *Mniodendron palmaeum* (Mitt.) Broth. in Engl. & Prantl, Nat. Pflanzenf. 1(3): 1172. 1909. — Synonymized by Touw (1971).

Mniodendron hellwigii Broth. in Warb., Bot. Jahrb. 16: 12 & 19. 1893. — Type: Papua New Guinea. Morobe, Finisterregebirge, 2 000–2 300 m, *Hellwig 317 p.p.* (H-BR!, holotype). — Synonymized by Touw (1971).

Mniodendron micholitzii Broth., Öfv. Finska Vet. Soc. Förh. 40: 192. 1898 ("*Micholitzii*"). — Type: Papua New Guinea. Milne Bay, Mountains near Mita, *Micholitz 131* (H-BR!, holotype). — Synonymized by Touw (1971).

Mniodendron densirameum Broth., Öfv. Finska Vet. Soc. Förh. 42: 120. 1900. — Type: Papua New Guinea. Central, Brown River Valley, *Musgrave s. n.* (H-BR!, holotype). — Synonymized to *M. hellwigii* Broth. by Dixon (1922).

Mniodendron kowaldii C. Müll., Hedwigia 41: 134. 1902 ("*Kowaldi*"). — Type: Papua New Guinea. Central, Mt. Yule, 7 000 ft., *Kowald 1297* (H!, lectotype, sel. Touw 1971). — Synonymized to *M. divaricatum* (Reinw.) Lindb. by Fleischer (1917).

Plants to 8 cm high, mostly umbellate but occasionally palmate, with branches to 4 cm long, green to dull green, not very glossy, erect to ascending with a well-defined stipe which has strongly reflexed leaves mostly hidden among the brown to rich red-brown rhizoids. Upper stipe leaves to 3.5 mm long, 1.6–2:1, not very keeled, distant, triangular to deltoid from a prominent auriculate base, often acuminate near apex. Ultimate branch leaves approximate to imbricated, not complanate, smooth to very lightly plicate when dry, erect to ascending, sometimes slightly secund, to 2.5 mm long, mostly about 4:1, broadest near the base and only gradually contracted to the acumen. Median laminal cells to 5 µm broad, mostly more than 10:1, usually smooth but occasionally slightly prorate, mostly rather thick-walled, with lumen/wall ratio 1–2:1. Alar cells in a poorly defined group, distinctly larger and more clear than the median laminal cells, to 25 µm broad, 1–2:1, forming distinctive, usually dentate auricles, especially on the stem leaves. Costa mostly slightly excurrent, dorsally smooth or with a few low teeth. Marginal cells, especially of branch leaves, often rather clear and green in 2–4 rows, shorter than interior cells with many of them only 1–2:1. Margins plane at base but somewhat down-flexed in distal portions of leaf, serrate nearly throughout, with teeth seldom composed of more than one cell, never with geminate teeth. Decurrency present on stem, often extending to next leaf, with the leaf margins sometimes confluent in plants with especially short internodes. Axillary hairs not offset from leaf insertion, to 120 µm long, with 2 basal brown cells and 1–2 terminal cells. Pseudoparaphyllia to 80 µm long, to 4 cells long and 2–3 cells wide at base. Macronematal apparatus present outside the decurrency on one side of nearly every stipe leaf, elliptical, to 50 µm wide, 3–4:1, to 3 cells wide and 4 cells long.

Dioicous. Seta to 4.5 cm long, red-brown, mostly at least somewhat flexed distally. Seta cross-section rounded pentagonal with three layers of yellow brown pachydermous cells, 1–2 layers of large, hyaline, leptodermous cells and a moderately well-defined central strand surrounded by small and pachydermous yellow brown cells. Urn to 3.5 mm, cylindric, about 4:1, mostly somewhat arcuate, strongly to only lightly sulcate at least in distal portions. Operculum inclined rostrate, often nearly as long as urn. Capsule wall mostly 3 layers thick. Exothecial cells in vertical rows of broad cells (to 20 µm wide) and narrow cells (to 12 µm wide), rather thin-walled, rectangular, mostly more than 4:1. Sto-

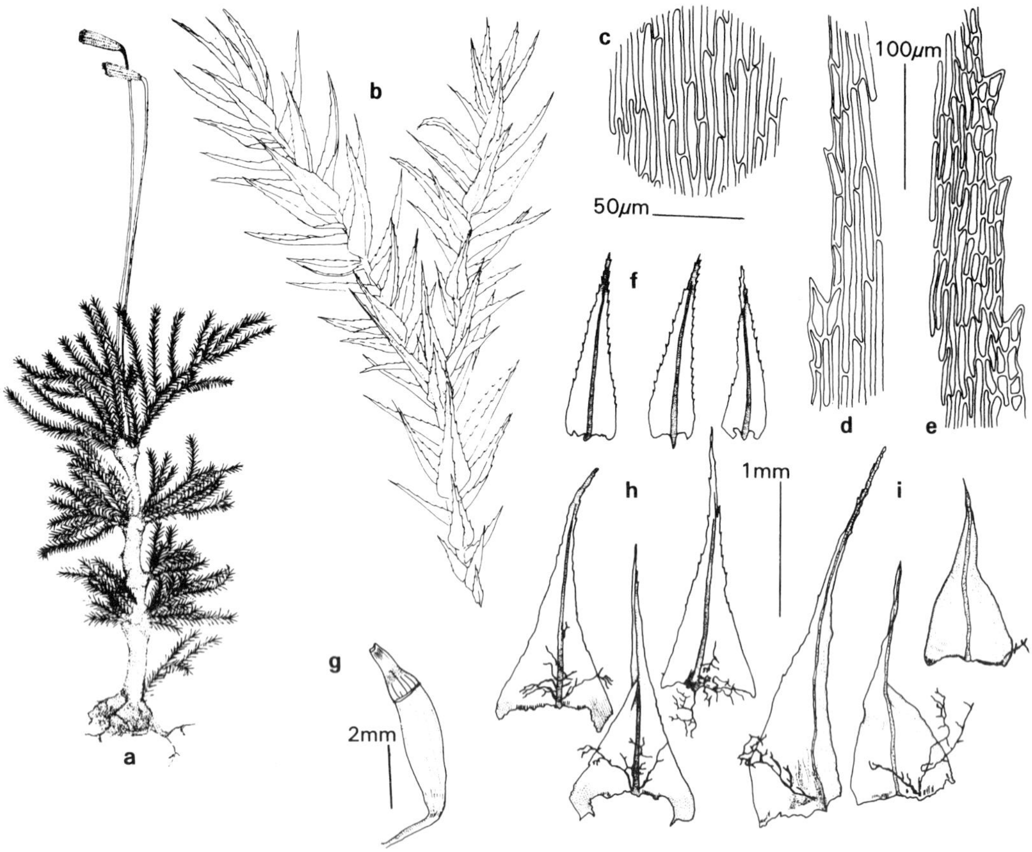

Fig. 4. *Hypnodendron dendroides* (Brid.) Touw (from *Creek 627*, H). — a: Growth habit. — b: Branch. — c: Leaf cells. — d: Leaf border of stipe leaf. — e: Leaf border of branch leaf. — f: Branch leaves. — g: Capsule. — h: Stipe leaves. — i: Perichaetial leaves. — Use the the 2 mm scale for h, the 50 μm scale for d, the 1 mm scale for g, i and j and the 100 μm scale for e and f.

mata superficial, restricted to short neck of capsule. Annulus vesicular, mostly not revoluble. Exostome to 800 μm long, dorsal plate more narrow than ventral one, horizontally closely striate, with a zig-zag median commisural line. Endostome sparsely papillose, nearly as long as the exostome, with basal membrane comprising 1/2–2/3 of its length, processes keeled and perforate, cilia 2–5. Spores smooth or nearly so, to 10 μm.

Illustrations: Dozy & Molkenboer 1865: tab. CCXXXIV (as *Mniodendron divaricatum*), tab. CCXXXV (as *M. humile*); Fleischer 1922: 1618, (fig. 255) as *M. divaricatum*; Touw 1971: 322 (fig. 37), 324 (fig. 38), 328 (fig. 39).

This is probably the most easily recognised of the New Guinea *Hypnodendron*. The combination of densely rhizoidal stipes and divergent stipe leaves should alert one to look for the auriculate bases with expanded, often isodiametric alar cells. It is exceptionally variable in such features as plant size and coloration, and the leaf shape, while usually rather narrow and auriculate, may vary in amount of acumination.

Hypnodendron dendroides was found at 900–3 300 m mainly in primary forests (39 localities). It was also found in disturbed forest (11), savannahs (4), gardens (2), scrub (1) and grassland (1). The substrates of our 140 specimens are: log or stump (77); tree, shrub or climber (44); soil (12); and boulder or outcrop (7).

Range on the Huon Peninsula: *1a*. 34598. *1b*. 34698. *1e*. 65021, 65035, 65098. *2b*. 33381, 33456. *2d*. 33662. *2e*. 33859. *2h*. 34055, 34075. *2i*. 34116, 34140. *2s*. 63936, 63937, 64000. *2t*. 64098, 64141. *2u*. 64177, 64242. *2w*. 64371, 64374, 64445. *2z*. 64818, 64819. *2A*. 65002. *3a*. 29239, 29378, 29387. *3b*. 29450. *3c*. 29576, 29581. *3f*.

29780. 29853, 29864. *3h.* 60128, 60290. *3k.* 60787, 60851. *3l.* 60966, 61002. *4b.* 583, 590, 608, 615, 627, 637, 712, 717, 723, 760. *4e.* 66458, 66485. *4g.* 66597, 66674, 66703, 66797. *4i.* 66984, 67005, 67029. *5c.* 65995. *6b.* 62591. *6l.* 32836. *6q.* 33041. *6s.* 33207, 33208. *6w.* 62825. *6z.* 63493. *6A.* 63586, 63620, 63666. *7b.* 65483, 65498, 65500. *7d.* 65740. *8a.* 30883, 30886, 31086. *8b.* 30974. *8d.* 31122, 31161, 31185. *8g.* 31374, 31375, 31401, 31409, 31420, 31422. *8h.* 31434. *9a.* 29949, 29961, 29996, 30237. *9b.* 30130, 30146. *9c.* 30192. *9d.* 61478, 61491, 61549, 61588, 61686. *9e.* 30410, 30422. *9f.* 30618. *9j.* 61167, 61248, 61298, 61328. *9k.* 61712, 61738. *9l.* 61974, 62049, 62129. *10a.* 28420, 28421, 28433. *10b.* 28483. *10c.* 28586, 28593, 28608. *10d.* 28691, 28728, 59518, 59568, 59646, 59658, 59668. *10e.* 28782, 28794. *10f.* 28866, 29024. *10i.* 29163. *10j.* 28997. *10n.* 59709, 59718, 59727. *10o.* 59794. *10p.* 59983, 60028, 60077.

Range in Western Melanesia: West Irian. *1–4, 6, 7.* Papua New Guinea. *9. 11.* Present report. *12.* Present report. *13–17, 20–23, 25, 27.* Solomon Islands. *31, 33.*

Total range: As 3: Sri Tha Vi; As 4: Ind Ma Phi PNG; Oc: Fij NC Sol Van.

4. *Hypnodendron diversifolium* Broth. & Geh. (Fig. 5)

in Broth., Öfv. Finska Vet. Soc. Förh. 40: 191. 1898. — Type: Papua New Guinea. Milne Bay, Mt. Dayman, up to 9 000 ft., *W. Armit junior 628* (H!, lectotype, *nov.*).

Hypnodendron macgregorii Broth. & Geh. in Broth., Öfv. Finska Vet. Soc. Förh. 42: 119. 1900 ("*Macgregorii*"). — Type: Papua New Guinea. 9 200 ft., *Mac Gregor s. n.* (H!, lectotype, *nov.*). —Synonymised by Touw (1971).

Mniodendron pygmaeum C. Müll., Hedwigia 41: 133. 1902. — *Hypnodendron pygmaeum* (C. Müll.) Broth., Nat. Pflanzenfam. 1(3): 1169. 1909. —Type: Papua New Guinea. Central, Moroka, 1 100 m, VIII.1893 *Loria s. n.* (destroyed in B). — Synonymised with reservations by Touw (1971).

Hypnodendron copelandii Broth., Öfv. Finska Vet. Soc. Förh. 47(14): 11. 1905 ("*Copelandii*"). — Synonymised by van Zanten (1964).

Plants to 7 cm high, mostly rather loosely umbellate but frequently palmate to pinnate, with branches to 3 cm long, green to pale green, mostly quite glossy, erect to ascending with a well-defined stipe which is usually nearly naked of rhizoids. Upper stipe leaves to 2.5 mm long, 1.6–2.0:1, usually so strongly reflexed that the distal 1/2 may be downwardly directed parallel to the stem, distally laterally twisted and canaliculate, distant, ovate-triangular, strongly constricted at mid-leaf into a long acumen. Lateral branch leaves approximate to imbricated, usually complanate, mostly not plicate when dry, loosely divergent but never secund, broadly ovate, to 2.0 mm long, about 1.6:1. Dorsal branch leaves appressed, similar to lateral ones but 1/2–3/4 as large and proportionately somewhat more narrow. Median laminal cells to 6 µm broad, mostly 6–10:1, smooth on stipe leaves but often lightly prorate on branch leaves, mostly rather thick-walled, with lumen/wall ratio 1–2:1. Alar cells not differentiated but basal cells distinctly larger and thicker-walled than median laminal cells, often orange in color. Costa mostly ending several cells below apex, dorsally with a several low to rather high teeth. Marginal cells mostly shorter than interior cells. Margins plane to base, dentate nearly throughout, with many teeth composed of more than one cell, geminate teeth absent. Leaves usually decurrent with that decurrency often extending to base of next leaf. Axillary hairs not offset from leaf insertion, to 100 µm long, with 2 basal brown cells and 1–2 terminal cells. Pseudoparaphyllia lanceolate and coarsely incised, to 80 µm long, to 4 cells long and 2–3 cells wide at base. Macronematal apparatus infrequent, not axillary, elliptical, to 50 µm wide, 2–3:1, to 3 cells wide and 5 cells long.

Dioicous. Seta to 4 cm long, yellow brown, mostly at least somewhat flexed distally. Seta cross-section round, with a well developed central strand surrounded by several layers of large (to 20 µm) hyaline and leptodermous cells which grade to 1–2 layers of yellow brown pachydermous and small (to 8 µm) external cells. Urn to 4 mm, cylindric, about 6:1, mostly somewhat arcuate, strongly to very lightly sulcate at least in distal portions. Operculum short conic rostrate, about 1/3 as long as urn. Capsule wall mostly 3 layers thick. Exothecial cells in vertical rows of broad cells (to 20 µm wide) and narrow cells (to 10 µm wide), rather thick walled, rectangular, 3–6:1. Stomata superficial, restricted to short neck of capsule. Annulus vesicular, mostly not revoluble. Exostome to 800 µm long, dorsal plate more narrow than the ventral one, horizontally closely striate, with a zig-zag median commisural line. Endostome sparsely papillose, nearly as long as the exostome, with basal membrane comprising 1/2–2/3 of its length, processes keeled and perforate, cilia 2–5. Spores smooth or nearly so, to 16 µm.

Illustrations: Touw 1971: 294, fig. 27.

Touw (1971) lists *Hypnodendron pygmaeum* (C. Müll.) Broth. as a probable synonym of *H. diversifolium* without having seen the original specimen. We also

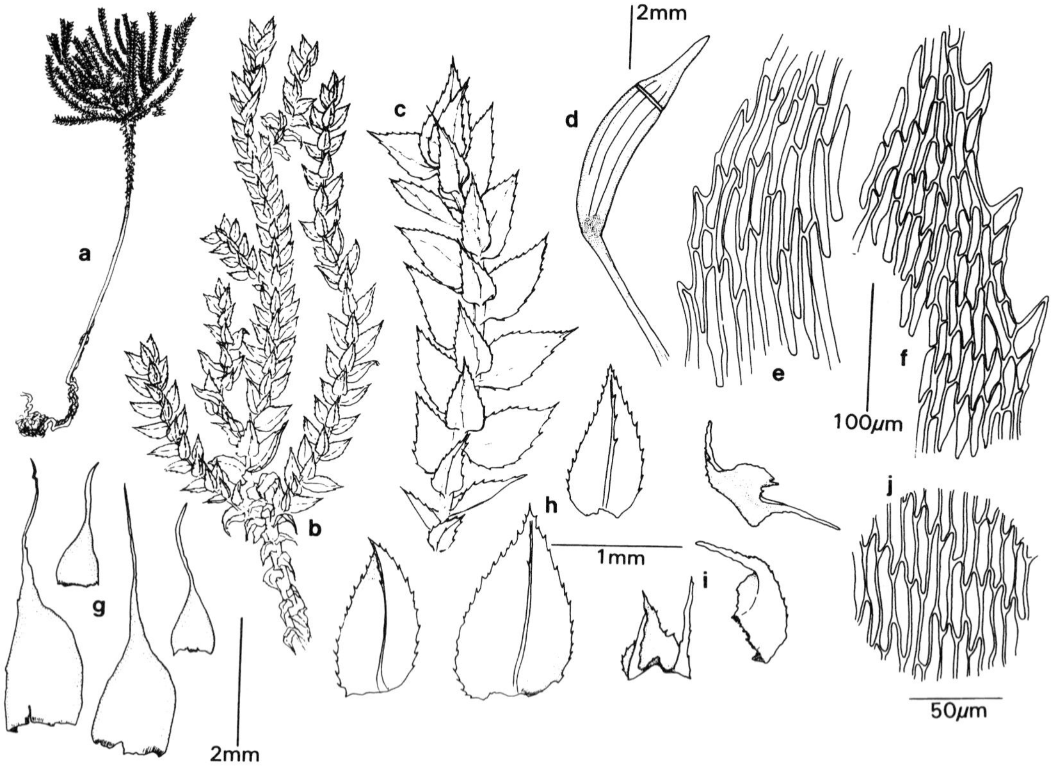

Fig. 5. *Hypnodendron diversifolium* Broth. & Geh. (a–c, j from *Norris 64068*, H; d–i from Koponen *28883*, H). — a: Growth form. — b, c: Close-ups of branches showing the tristichous leaf arrangement. — d: Capsule. — e: Margin at mid-shoulder of stipe leaf. — f: Leaf border of branch leaf. — g: Perichaetial leaves. — h: Branch leaves. — i: Stipe leaves. — j: Leaf cells. — Use the 2 mm scale on bottom for g, the 2 mm scale on top for d, the 100 µm scale for e and f, the 1 mm scale for h and i and the 50 µm scale for j.

have not been able to find that specimen and so we can only continue Touw's indication of probable synonymy.

The tristichous and dimorphous branch leaves of this plant are the basis for the sectional name *Tristichophyllum* proposed by Touw (1971). This feature should not be overemphasized as it can occasionally be somewhat obscure. Even at best development the smaller dorsal leaves are seldom less than 1/2 of the length of the lateral leaves, and frequently the difference in size is much less. It may further be noted that such leaf dimorphism of the complanate stems may be approached in *Hypnodendron subspinervium*, *H. reinwardtii*, and *H. vitiense*. A better feature, allowing identification of nearly all specimens, is the constriction at the middle of the stipe leaves which defines a long, laterally flexed, downwardly directed acumen.

Hypnodendron diversifolium was found at 750–3 500 m mainly in primeval rainforest (23 localities). It was also found in second growth forest (3), garden (1) and scrub (1). The substrates of our 45 specimens are: soil and humus (20), tree or shrub (15), log (8), and boulder (2).

Range on the Huon Peninsula: *1a*. 34576. *2t*. 64068. *2z*. 64882, 64887. *3a*. 29288, 29334, 29401. *3f*. 29852. *3h*. 60333. *4b*. 647, 772. *4c*. 794. 4i. 67025. *5c*. 66108. *6h*. 32476. *6w*. 62710B. *6A*. 63699, 63779, 63783. *8d*. 31158. *8h*. 31439, 31498. *9a*. 29958, 29999. *9d*. 61475, 61477, 61572, 61581, 61666, 61675. *9e*. 30430, 30460. *9j*. 61214, 61250. *9k*. 61827, 61888. *9l*. 62036. *10d*. 59590. *10e*. 28829. *10f*. 28883. *10k*. 59308. *10o*. 59799, 59849. *10p*. 60084. *14e*. 58980.

Range in Western Melanesia: West Irian. *2*, *7*. Papua New Guinea. *9*. Star Mts. between Folongonom and Artis Camp, 2 500–2 600 m, terrestrial, *Touw 17976* (H). *11*. Present report. *12*. Present report, and: Slate Creek and Gumi

Creek Divide, 2 100 m, on log and tree, *Streimann 13864, 13909, 13929* (H). *13, 14.*: Kum Magei Mtn. 2 550 m, on gound, *Streimann 20698* (H). *16.* Near Hogabi Village, 1 850 m, on ground and tree roots and stumps, *Streimann 18636, 18646, 18663* (H). *17, 20, 22, 28.*

Total range: As 3: Vi; As 4: Ind Ma Phi PNG; Oc: Sol.

5. *Hypnodendron fuscomucronatum* (C. Müll.) Jaeg. (Figs. 6, 12)

in Jaeg. & Sauerb., Ber. Thät. St. Gall. Naturw. Ges. 1877/78: 360. 1879 (Adumbratio 2: 624). — *Hypnum fuscomucronatum* C. Müll., Bot. Zeit. 20: 393. 1862. — *Mniodendron fuscomucronatum* (C. Müll.) Broth., Öfv. Finska Vet. Soc. Förh. 47: 12. 1905.

Hypnodendron chalmersii Mitt., Proc. Linn. Soc. New South Wales 7: 103. 1882 ("*Chalmersii*"). — *H. fuscomucronatum* subsp. *chalmersii* (Mitt.) Touw, Blumea 19: 337. 1971. — Type: Papua New Guinea. Central, Spur of the Owen Stanley Range, *J. Chalmers s. n.* (NY, holotype, not seen, isotype in H!). — Reduced to subspecific level by Touw (1971).

Hypnodendron fusco-aciculare C. Müll. in Geh., Biblioth. Bot. 13: 10, *Taf. 8, figs. 1–16.* 1889. — Type: Papua New Guinea. Milne Bay, Cloudy Mts. near South Cape, *Chalmers & Bridge s. n.* (destroyed in B, not seen). — Synonymized by Dixon (1922).

Hypnodendron brassii Bartr., Contrib. U. S. Nat. Herb. 37: 52. 1965. — Type: Papua New Guinea. Eastern Highlands, Arau, 1 400 m, *Brass 32042* (holotype in US or FH, see below, not seen). — Synonymized by Touw (1971).

Plants to 10 cm high, mostly umbellate, with branches to 5 cm long, fuscous-green, not glossy, usually erect, with a well-defined densely tomentose stipe which has imbricated, appressed leaves. Leaves approximate to imbricated, symmetric, flaccid and thus appearing irregularly plicate when dry, straight or only slightly laterally twisted, canaliculate-keeled, often strongly so, and thus appearing somewhat concave at base. Upper stipe leaves symmetric, to 4 mm long, mostly about 3:1, deltoid to deltoid-lanceolate, broadest immediately above their bases, short acuminate with the acumen somewhat divergent when dry. Ultimate branch leaves not complanate, erect to ascending, deltoid to oblong-ovate, to 2 mm long, mostly about 2.5 1, cordate but not auriculate, broadest near their base and gradually contracted to a very short acumen, usually with a yellowish to red-brown terminal cell. Median laminal cells to 6 µm broad, mostly less than 8:1 in branch leaves but longer in the stipe leaves, smooth to lightly prorate, mostly rather thick-walled, with lumen/wall ratio 1–2:1. Alar cells mostly in a well defined group, green and thin-walled, often quadrate, to 18 µm broad; occasionally rather poorly differentiated in the stipe leaves. Basal cells enlarged and thicker-walled relative to median laminal cells, usually concolorous. Costa usually without dorsal teeth, mostly filling the short acumen but occasionally percurrent. Marginal cells shorter than interior cells with many of them only 1–2:1. Margins plane throughout, dentate in distal 3/4, mostly with small teeth composed of one cell or only a fraction of a cell; geminate teeth absent. Leaves decurrent but with decurrency not extending to next leaf. Axillary hairs not offset from leaf insertion, to 70 µm long, with 2 basal brown cells and 1–2 terminal cells. Pseudoparaphyllia linear, to 80 µm long, about 5 cells long and 2 cells wide at base. Macronematal apparatus not axillary, frequent on the outside of the decurrency of the stipe leaves, short elliptical, to 100 µm wide, about 1.5:1, 5 cells in width and length.

Dioicous. Seta to 3 cm long, brown to red-brown, distally geniculate. Seta cross-section rounded-triangular with two layers of small pachydermous cells and several layers of abruptly larger, leptodermous interior cortical cells, and with a few pachydermous cells around the central strand. Urn to 6 mm, cylindric, about 6:1, mostly arcuate and sulcate. Operculum rostrate, about 1/2 as long as urn. Capsule wall to 120 µm thick with 4–5 layers of cells. Exothecial cells rectangular, 3–8:1, in well defined vertical bands of broad (to 20 µm) cells and narrower (to 10 µm cells); little differentiated at mouth into isodiametric cells. Stomata superficial, restricted to short neck of capsule. Annulus vesicular, mostly not revoluble. Exostome to 800 µm long, dorsal plate more narrow than ventral one, horizontally closely striate, with a zig-zag median commisural line. Endostome sparsely papillose, about as long as the exostome, with basal membrane comprising 1/2–2/3 of its length, processes keeled and perforate, cilia 2–5. Spores smooth or nearly so, to 15 µm.

Illustrations: Geheeb 1889: taf. VIII (as *H. fusco-aciculare)*; Touw 1971: 334 (fig. 40).

Touw (1971) indicated the holotype of *Hypnodendron brassii* to be in FH. According to the protologue the type is in U.S. National Herbarium. However, see Bartram (1965), the footnote on p. 43.

Hypnodendron fuscomucronatum resembles *H. auricomum* in having leaves appressed and rather

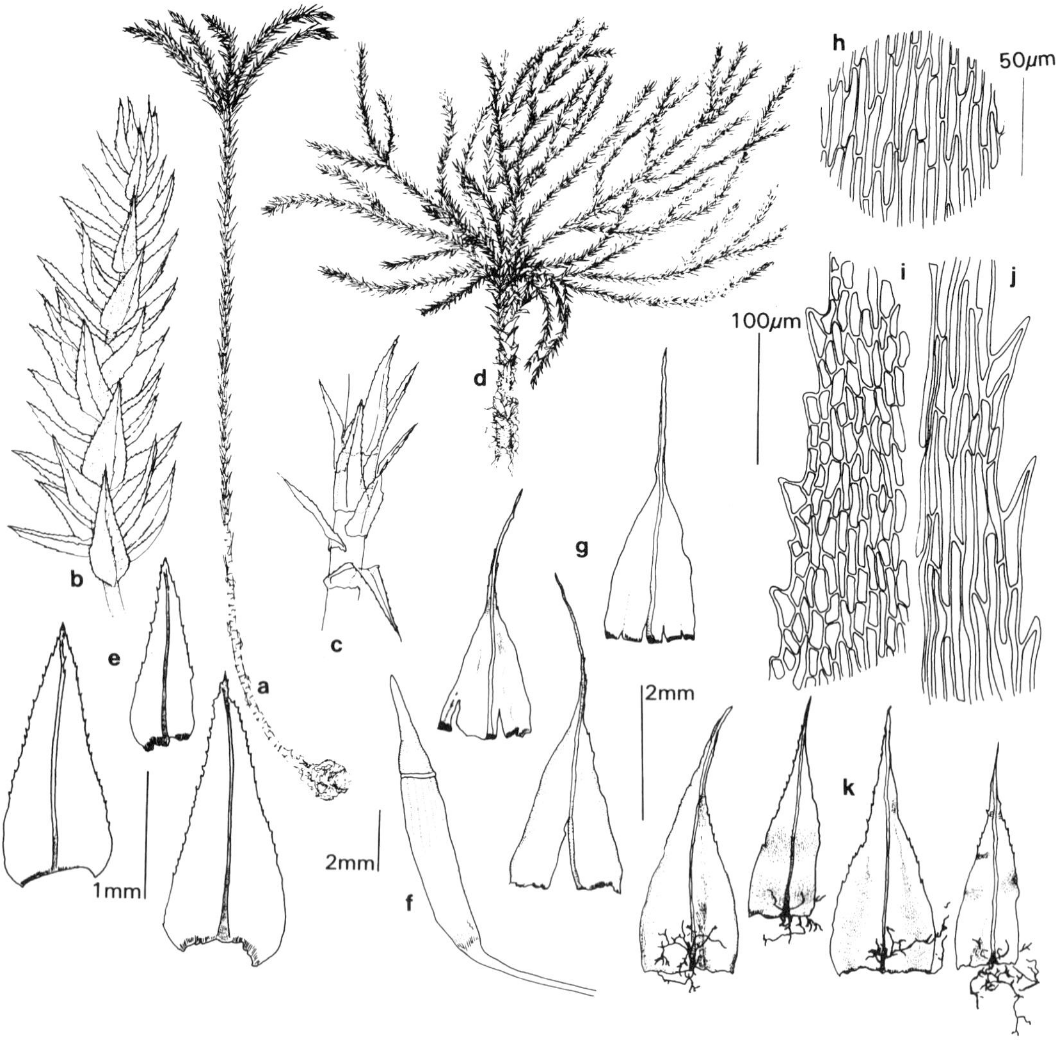

Fig. 6. *Hypnodendron fuscomucronatum* (C. Müll.) Jaeg. (a–d) from *Norris 63795*, UC; e, h–j from *Norris 63795*, H, f and g from *Koponen 28626*, H). — a: Growth habit. — b: Branch. — c: Detail at mid-stipe. — d: Frond and apex of stipe. — e: Branch leaves. — f: Capsule. — g: Perichaetial leaves. — h: Leaf cells. — i: Leaf border of branch leaf. — j: Leaf border of stipe leaf. — k: Stipe leaves. — Use the 1 mm scale for e, the 2 mm scale on the left for f, the 2 mm scale on the right for g and k, the 50 μm scale for h and the 100 μm scale for i and j.

densely covering most of the stipe. We discuss its differentiation from *H. auricomum* under that species. From the two other local species with appressed leaves, *H. fuscomucronatum* is distinctive in its very small, never geminate teeth, and in the enlarged, often thin-walled alar cells of its branch leaves.

Touw recognises our material under *Hypnodendron fuscomucronatum* subsp. *chalmersii*.

Hypnodendron fuscomucronatum was found at 750–2 100 m mainly in primeval rainforest (6 localities). It was also found in second growth forest (1), and savannahs (1). The substrates of our 12 specimens are: soil (6), tree or shrub (3), log (2), and boulder (1).

Range on the Huon Peninsula: *4e*. 66466. *6q*. 33057. *6A*. 63794, 63795. *7d*. 65607, 65695. *10c*. 28582, 28626, 28641. *10d*. 28714. *10j*. 28917. *14e*. 58997.

Range in Western Melanesia: West Irian. *1*, *2*. Papua New Guinea. *10*, *11*. *12*. Present report, and: Upper Watut Logging Area, 1 720 m, on rock, *Streimann 33067* (H). *14*. Mt. Hagen, 2 000 m, on stone in creek, *Vink 16796A* (H). *20*. K. B. Sawmill, 600 m, on rocks, *Streimann & Naoni 16506* (H). *22*. Mt. Suckling 9.34S × 148.56E, 1 775 m, banks of rivulet, *Veldkamp & Stevens 5792* (H). *24.* Solomon Islands. *31.*

Total range: As 4: Ind Phi PNG; Oc: Sol Van.

6. *Hypnodendron junghuhnii* (C. Müll.) Jaeg. (Fig. 7)

in Jaeg. & Sauerb., Ber. Thät. St. Gall. Naturw. Ges. 1877/78: 358. 1879 (Adumbratio 2: 622). — *Hypnum junghuhnii* C. Müll., Syn. Musc. Frond. 2: 506. 1851 ("*Junghuhnii*").

Plants to 15 cm long, mostly palmate to nearly umbellate with branches to 4 cm long, glossy green to pale green. Stipe well-defined with rhizoids restricted to the base, and with leaves patent, distant, deltoid triangular gradually narrowed to apex or somewhat acuminate, to 2.5 mm long, 1.6–2:1, symmetric and not keeled. Ultimate branch leaves ovate to ovate lanceolate, to 3.5 mm long, about 2.5:1, broadest near the middle, approximate to imbricated, loosely spreading, not strongly keeled. Median laminal cells to 7 µm broad, mostly more than 10:1, smooth to minutely prorate, thick-walled and porose. Alar cells indistinctly differentiated, to 12 µm broad, 2–4:1, thick-walled. Interior basal cells thicker walled and more pitted than alar cells. Costa percurrent to subpercurrent, mostly with several dorsal teeth. Margins mostly plane, dentate with some teeth, especially on the branch leaves, geminate, and with some teeth multicellular. Axillary hairs only slightly offset from leaf insertion, very small, to 3 cells long with 2 basal brown cells and a globular terminal cell to 40 µm long. Stems round in cross section, with a well developed central strand surrounded by large and leptodermous inner cortical cells which grade into several layers of small exterior cells. Pseudoparaphyllia linear, to 100 µm long, to 6 cells long and 2 cells wide at base. Macronematal apparatus not axillary, to 50 µm wide, about 3:1, about 2–3 cells wide by 3–4 cells long.

Dioicous. Seta 3–5 cm long, brown to reddish brown, mostly strongly flexed distally. Urn to 6 mm, cylindric, 4–6:1, mostly straight above the arc of the seta, strongly sulcate. Operculum conic, to 1/3 of urn length. Sporangial wall to four cell layers thick, with cells thick-walled, irregular in size and shape. Exothecial cells to 40 µm wide, thick-walled, irregularly rectangular, 1.2–2:1. Exostome to 800 µm, dorsal plate more narrow than the ventral one, horizontally closely striate, with a zig-zag median commisural line. Endostome sparsely papillose, about as long as the exostome, with basal membrane comprising 1/2–2/3 of its length, processes keeled and perforate, cilia 3–5. Spores smooth or nearly so, to 15 µm.

Illustrations: Dozy & Molkenboer 1855–1870: 1612 (fig. 254); Fleischer 1922: 1612 (fig. 254); Touw 1971: 287 (fig. 25).

This is one of the more easily identified of the local *Hypnodendron*. It has quite distantly inserted deltoid stipe leaves without a tendency for the upper leaves to flex from the stem upon drying.

Range on the Huon Peninsula: *Hypnodendron junghuhnii* was not among our Huon Peninsula collections.

Range in Western Melanesia: Papua New Guinea. *12. 22.*

Total range: As 3: Tha; As 4: Ind Ma Phi PNG.

7. *Hypnodendron menziesii* (Hook.) Par. (Fig. 7)

Ind. Bryol. 604. 1896. — *Hypnum menziesii* Hook., Musci Exot. 1: 33. 1818. — *Sciadocladus menziesii* (Hook.) Jaeg. ex Broth., Nat. Pflanzenfam. 1(3): 1168. 1909.

Hypnodendron splendidum Besch., Ann. Sci. Nat. Bot. Sér. 5, 18: 245. 1873. — *Sciadocladus splendidus* (Besch.) Jaeg. ex Broth., Nat. Pflanzenfam. 1(3): 1168. 1909. — *Hypnodendron menziesii* subsp. *splendidum* (Besch.) Touw, Blumea 19: 265. 1971. — Reduced to subspecific status under *Hypnodendron menziesii* by Touw (1971).

Plants to 15 cm high, umbellate to palmate, with branches to 3 cm long, glossy yellow green to green, usually erect, with a well-defined stipe which has spreading to ascending leaves and has few rhizoids except at its base. Stipe leaves readily lost, symmetric, to 3.5 mm long, mostly 2–2.5:1, not keeled, ovate but contracted to an acumen which occupies about 1/2 of the leaf. Ultimate branch leaves approximate to imbricated, symmetric, pentastichous but sometimes obscurely so, not rugose or plicate, erect to ascending, to 2.5 mm long, mostly about 2:1, broadest about 1/3 from base, rather abruptly contracted near apex into the short acumen. Median laminal cells to 6 µm broad, mostly more than 10:1, essentially smooth to lightly prorate, mostly rather thick-walled, with lumen/wall ratio 2–3:1. Alar cells well differentiated, to 10 µm wide, 1–2:1, thick-walled with lumen/wall ratio about 1:1, usually orange to red-brown. Costa with large dorsal teeth, excurrent with excurrency terete at its base, to 0.25 mm long. Margins recurved near base. Branch leaves coarsely dentate in distal 1/2 but nearly entire below; stipe leaves virtually entire except near apex. Geminate

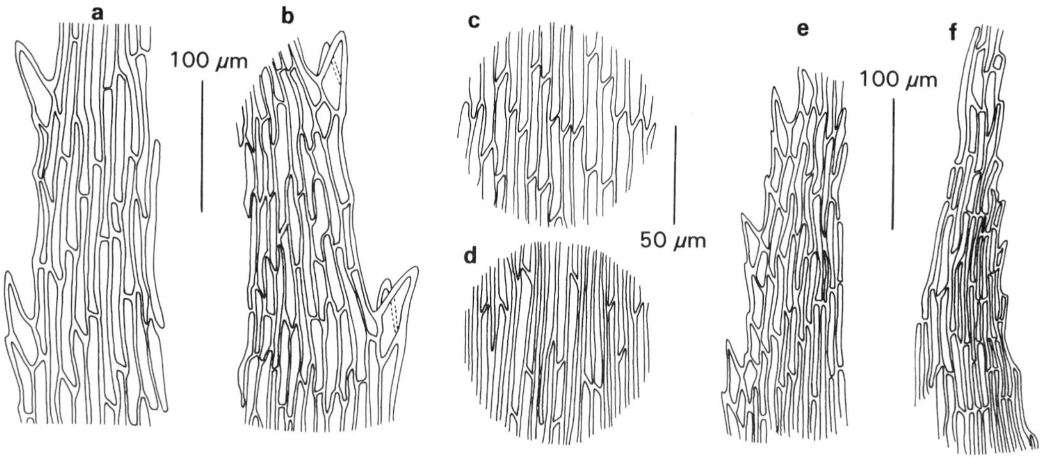

Fig. 7. — a–c. *Hypnodendron junghuhnii* (C.Müll.) Jaeg. (from Fleischer "Kryptogamae exsiccatae 2000", H): — a: Leaf border of stipe leaf. — b: Leaf border of branch leaf. — c: Leaf cells. — d–f. *H. menziesii* (Hook.) Par. (from *Le Rat s.n.*, H-BR "342"). — d: Leaf cells. — e: Leaf border of branch leaf. — f: Leaf border of stipe leaf. — Use the 50 μm scale for c and d and the 100 μm scale for a, b, e, and f.

teeth absent. Leaves, especially on stipe, short decurrent with decurrency not extending to next leaf. Axillary hairs not offset from leaf insertion, to 40 μm long, with 1 basal brown cell and a single terminal cell. Pseudoparaphyllia lanceolate, entire, to 100 μm. Macronematal apparatus scattered and very infrequent, orbicular, to 50 μm wide, nearly isodiametric, to 2 cells in each dimension. Stem nearly round in cross-section, with a strong central strand surrounded by moderately thick-walled, pale inner cortical cells to 20 μm in diameter, and with an abruptly defined group of several layers of very thick walled outer cortical cells to 8 μm in diameter.

Dioicous. Seta to 4 cm long, red-brown, straight to somewhat flexed distally. Seta cross-section round to elliptic with 2–4 layers of brown pachydermous cells grading imperceptibly into the white, rather thin-walled inner cells which surround the central strand. Urn to 6 mm, cylindric, about 5:1, mostly not arcuate, never sulcate. Operculum conic, to 1/4 as long as urn. Capsule wall to 120 μm thick with only a single layer of cells which are isodiametric in cross-sectional view and equally thickened on all visible walls. Exothecial cells very irregular in size and shape, mostly rectangular, 1–4:1, to 45 μm broad, not in organized streaks of broad and narrow cells. Stomata superficial, restricted to short neck of capsule. Annulus vesicular, mostly not revoluble. Exostome to 500 μm long, dorsal plate more narrow than the ventral one, horizontally closely striate, with a zig-zag median commisural line. Endostome nearly smooth, about as long as the exostome, with basal membrane comprising 1/2–2/3 of its length, processes keeled and perforate, cilia 2–4. Spores smooth or nearly so, to 25 μm.

Illustrations: Sainsbury 1955: (pl. 47, fig. 2 as *Sciadocladus menziesii*); Touw 1971: 262 (fig. 15).

This plant has many highly unusual features within the genus. In the introductory paragraph on the genus, we have already referred to the single cell layer of the capsule wall. This should be more fully explained. In nearly all mosses, the developing capsule wall is multicellular. Late in development certain of these cells typically deposit thicker walls and persist as the jacket of the mature sporocarp. The other layers of the capsule wall collapse and are partially resorbed as the capsule matures. Our observation of multicellular walls in most members of the genus *Hypnodendron* is based upon observation of mature capsule walls after the collapse of the interior layers.

We have not studied the only Western Melanesian specimen of *Hypnodendron menziesii* (see below), and our studies and drawings of the plant come from a New Caledonia example in H-BR (Mt. Mon, January 1906, *LeRat s.n.*). Touw (1971) cites the above specimen and places it, with the Solomon Island collection cited above, in subsp. *splendidum* (Besch.) Touw. In accord with our usual practice, we are not here recognising taxa at infra-specific levels.

Range on the Huon Peninsula: *Hypnodendron menziesii* was not among our Huon Peninsula collections.

Range in Western Melanesia: Solomon Islands. *31*. Mt. Popomanasiu, *van Zanten 682617* (GRO, L; Touw 1971).

Total range: Oc: NC Sol; Austr 2.

8. *Hypnodendron milnei* Mitt. (Figs. 8, 12)

in Seemann, Flora Vitiensis 401. 1873 ("*Milnei*"). — *Mniodendron milnei* (Mitt.) Mitt., Proc. Linn. Soc. N. S. Wales 7: 103. 1882.

Mniodendron parvum C. Müll., Hedwigia 41: 133. 1902. — *Hypnodendron parvum* (C. Müll.) Dix., J. Linn. Soc. Bot. 45: 494. 1922. — *H. milnei* subsp. *parvum* (C. Müll.) Touw, Blumea 19: 301. 1971. — Type: Papua New Guinea. Central, Moroka, 1 300 m, *Loria 1638* (BM, lectotype, sel. by Touw 1971, not seen). — Reduced to subspecific level under *H. milnei* by Touw (1971).

Mniodendron milnei (Mitt.) Par. fo. *papuana* Fleisch., Engl. Bot. Jahrb. 55: 35. 1917. — Type: Papua New Guinea. Felsspitze im buschwaldähnlichen Gebirgswald mit viel Epiphyten und Lichtungen mit felsigem, nassem Gelände auf schattigem Boden, 1 400–1 500 m, 14.VIII.1913 *Ledermann 12886* (not seen). — Synonymized by Touw (1971) with *Hypnodendron milnei* subsp. *parvum*.

Mniodendron korthalsii Bosch & Sande Lac., Bryol. Jav. 2: 139, *tab. 236*. 1865 ("*Korthalsii*"). — *Hypnodendron milnei* subsp. *korthalsii* (Bosch & Sande Lac.) Touw, Blumea 19: 302. 1971. — Reduced to subspecific level under *Hypnodendron milnei* by Touw (1971).

Plants green to fuscous green, erect to ascending with a well-defined stipe with rhizoids restricted to the base, to 10 cm high, palmate to pinnately branched with branches to 4.5 cm long. Stipe leaves to 3 mm long, 2–2.5:1, symmetric, closely inserted, spreading to reflexed from a clasping base which occupies up to 1/3 of the leaf, lanceolate from a rounded to almost auriculate insertion, gradually narrowed from the top of the clasping base to a narrow acumen. Ultimate branch leaves overlapping, loosely spreading, plicate but not keeled, to 2.5 mm long, 3–5:1, broadest near the base. Median laminal cells to 7μm broad, about 6–15:1, smooth to very low prorate, rather thick-walled, often pitted. Surface cells of the subula of the stipe leaves incrassate, to 6:1, about 6 μm wide. Laminal cells in subula often short, sometimes isodiametric. Basal cells pellucid, about 10:1, to 12 μm broad in a well defined area throughout the sheathing base. Alar cells well differentiated, to 25 μm broad, 1–3:1, thick-walled and pitted, often orange-brown. Costa excurrent, usually shortly so, with a few low dorsal teeth. Margins basally slightly recurved, entire on the clasping base of the stipe leaves but elsewhere multicellular dentate with sharp, brown-tipped teeth; many teeth geminate. Marginal cells between the teeth mostly abruptly shorter than adjacent interior cells, forming a moderately well defined border. Axillary hairs to 80 μm long, not offset from leaf insertion, to 4 cells long with 2 basal brown cells and a pair of longer terminal cells. Stems rounded quadrangular with a strong central strand surrounded by hyaline and leptodermous inner cortical cells to 20 μm in diameter, and with several abruptly defined, layers of very thick-walled outer cortical cells to 8 μm in diameter. Pseudoparaphyllia linear, to 50 μm long, 3–4 cells long and 2 cells wide at base. Macronematal apparatus with all cells elongate, nearly axillary on the outside of an imaginary line of decurrency from some of the stipe leaves, elliptical, to 100 μm wide, about 3:1, to 3 cells wide and 4 cells long.

Dioicous. Seta 3–5 cm long, brown to reddish brown, mostly at least somewhat flexed distally. Seta cross-section rounded pentagonal with three layers of yellow brown pachydermous cells, 1–2 layers of large, hyaline, leptodermous cells and a moderately well-defined central strand surrounded by small and pachydermous yellow brown cells. Urn to 6 mm, cylindric, 4–6:1, mostly somewhat arcuate, usually lightly sulcate at least in distal portions. Operculum rostrate, to 1/2 of urn length. Sporangial wall to 5 cell layers (150 μm) thick-with cells thick walled and rather uniform throughout. Exothecial cells to 30 μm wide, thick walled, elongate hexagonal, 2–3:1. Exostome to 600 μm long, dorsal plate more narrow than the ventral one, horizontally closely striate, with a zig-zag median commisural line. Endostome sparsely papillose, nearly as long as the exostome, with basal membrane comprising 1/2–2/3 of its length, processes keeled and perforate, cilia 2–5. Spores smooth or nearly so, to 15 μm.

Illustrations: Dozy & Molkenboer 1865: tab. CCXXXVI (as *Mniodendron korthalsii*); Touw 1971: 298 (fig. 29), 300 (fig. 30).

Touw (1971) recognises three subspecies of *Hypnodendron milnei*. Two of these occur in Western Melanesia, and the third occupies most of Indonesia and the Philippines. In this Huon Peninsula series we have followed a policy of recognising taxa only at the specific level and higher. We will continue this policy here.

Our Western Melanesian specimens of *Hypnodendron milnei* are quite easily recognised by the

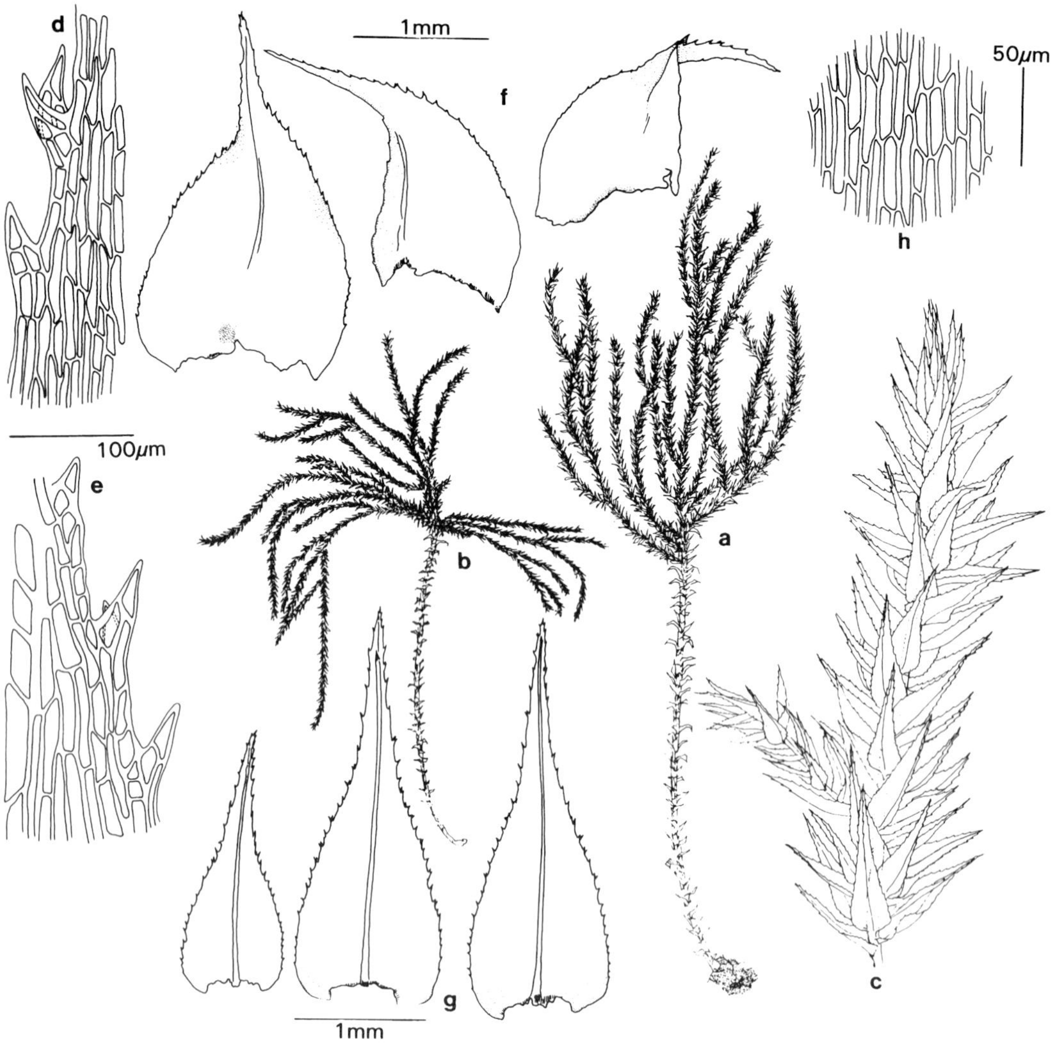

Fig. 8. *Hypnodendron milnei* Mitt. (a–c from *Norris 61611*, UC; d–h from *Norris 61613*, H)). — a, b: Growth habit. — c: Branch. — d: Leaf border of branch leaf. — e: Leaf border of stipe leaf. — e: Branch leaves. — f: Stipe leaves. — g: Branch leaves. — h: Leaf cells. — Use the100 μm scale for d and e, the 1 mm scale on top for f, the 1 mm scale on bottom for g and the 50 μm scale for h.

pattern of clasping leaf bases. No other local member of the genus shows such a character. Unfortunately, many of the variations of this widely distributed species may rather poorly exhibit the clasping leaf bases. If the clasping base were to be ignored, *H. milnei* would go to the *Phoenicobryum* group. It is easily distinguished from those species by the lack of any complanation of its leaves, and by the fuscous color which reminds one of *Hypnodendron fuscomucronatum*. Even when the clasping base is not noted, *Hypnodendron milnei* can be identified by the loose areolation of its leaf base with large and pellucid cells (especially well observed at the margin near the junction of the base and the limb).

Hypnodendron milnei was found at 1 800–2 230 m mainly in primeval rainforest (4 localities). It was twice found in a second growth forest. The substrates of our 6 specimens are: soil (5) and tree (1).

Range on the Huon Peninsula: *3f.* 29867. *3k.* 60778, 60889. *9d.* 61611, 61613. *10b.* 28509, 28510.

Range in Western Melanesia: West Irian. *2*, *4*, *6*, *7*.

Papua New Guinea. *9*. Folongonom, moist rocks, 2 300 m, *Touw 17477* (H), Busilmin, terrestrial, 1 500–1 550 m, *Touw 18484* (H). 12. Present report. *20*, *23*. Solomon Islands. *31*.
Total range: As 4: Ind Ma Phi PNG; Oc: Sol Van.

9. *Hypnodendron reinwardtii* (Schwaegr.) Jaeg. (Figs. 9, 12)

in Jaeg. & Sauerb., Ber. Thät. St. Gall. Naturw. Ges. 1877/78: 358. 1879 (Adumbratio 2: 622). — *Hypnum reinwardtii* Schwaegr., Species Musc. Suppl. 3, 1: tab. 223, *figs. 1–16*. 1828 ("*Reinwardti*").

Hypnodendron caducifolium Herz., Hedwigia 61: 292. 1919. — *H. reinwardtii* subsp. *caducifolium* (Herz.) Touw, Blumea 19: 245. 1971. — Reduced to subspecific status under *H. reinwardtii* by Touw (1971).

Plants stramineous to rust red but light green in younger portions, often quite glossy, erect to ascending, usually with a well-defined stipe which has squarrose leaves and has rhizoids restricted to its base, to 10 cm high, pinnately to weakly bipinnately branched with branches to 2.5 cm long, often attenuate and microphyllous distally. Upper stipe leaves approximate, ovate lanceolate, gradually narrowed to an acumen which occupies about 1/3–1/5 of the leaf, to 3 mm long, 2–3.5:1, asymmetric, so keeled basally as to resist lying flat on a slide, usually plicate on each side of the keel. Ultimate branch leaves spreading to ascending, overlapping, to 2.5 mm long, 3–5:1, broadest at about 1/4 from the base, less strongly acuminate than the stipe leaves, keeled at base. Median laminal cells to 5 µm broad, mostly more than 10:1, smooth, often with red discoloration of internal contents, thick walled with lumen/wall ratio 0.5–1:1. Alar and basal cells to 18 µm broad, 2–4:1, thick-walled and pitted with lumen/wall ratio about 11, orange to red. Interior basal cells thicker walled and more pitted than alar cells. Costa excurrent, usually with only a few low dorsal teeth. Margins basally slightly recurved, especially on stipe leaves, dentate with multicellular teeth, geminate teeth scarce, usually with short cells on margin between the teeth. Axillary hairs to 120 µm long, only slightly offset from leaf insertion, to 4 cells long with 2 basal brown cells and two similar terminal ones. Stems rounded pentagonal in cross section, with a strong central strand surrounded by large and leptodermous inner cortical cells to 25 µm which gradually grade into the very thick-walled, reddish-brown outer cortical cells. Pseudoparaphyllia scarce, to 80 µm long, lanceolate and erose. Macronematal apparatus present in an apparently axillary position outside the imaginary decurrent line from one side of nearly every stipe leaf, elliptical, to 300 µm wide, 1–1.5:1, up to 15 cells wide.

Dioicous. Seta to 8 cm long, brown to reddish brown, mostly straight to apex. Seta cross-section round, with 2–3 layers of yellow brown pachydermous cells, 1–2 layers of large, hyaline, leptodermous cells and a moderately well-defined central strand. Urn to 8 mm, cylindric, 4–8:1, mostly straight, lightly to rather strongly sulcate. Operculum rostrate, nearly as long as the urn. Capsule wall 4–5 cell layers thick with all cells thick-walled and similar. Exothecial cells to 20 µm wide, thick-walled, irregularly rectangular, 1–2:1, alternating with streaks of cells to 10 µm wide and 3–6:1. Exostome to 800 µm long, dorsal plate more narrow than the ventral one, horizontally closely striate, with a zigzag median commisural line. Endostome sparsely papillose, nearly as long as the exostome, with basal membrane comprising 1/2–2/3 of its length, processes keeled and perforate, cilia 2–4. Spores smooth or nearly so, to 25 µm.

Illustrations: Dozy & Molkenboer 1855–1870: tab. CCXXXIII; Bartram 1939: pl. 12, (fig. 188); Fleischer 1922: 1608 (fig. 253); Touw 1971: 215 (pl. Ie); 240 (fig. 6a, d–h, m–o), 242 (fig. 7 a, b, e, k–m).

Hypnodendron reinwardtii is rather commonly collected throughout its rather wide geographic range. The specimens from these collections are, however, often very scanty because of its frequent habitat on small twigs. Even within our quite limited collection, we have found it quite variable. In our key we emphasize the length/width ratio of the leaves. Upper stipe leaves and adjacent leaves of larger branches should be the primary area of focus for these measurements. The estimate of the length/width ratios of these leaves should not be made by simple casual inspection of the dry plant. Many of these dry plants have leaves so shrunken by drying that they will give the illusion of being very narrow.

Hypnodendron reinwardtii was found at 1 450–3 100 m mainly in primeval rainforest (6 localities). It was once found in grassland. The substrates of our 7 specimens were consistently epiphytic.

Range on the Huon Peninsula: *1e*. 65100. *2v*. 64190. *9d*. 61542, 61593. *9f*. 30583A. *9k*. 61752. *10c*. 28631.

Range in Western Melanesia: West Irian, *7*. Papua New Guinea, *9*. Above Folongonom, on twigs, 2 420 m, *Touw*

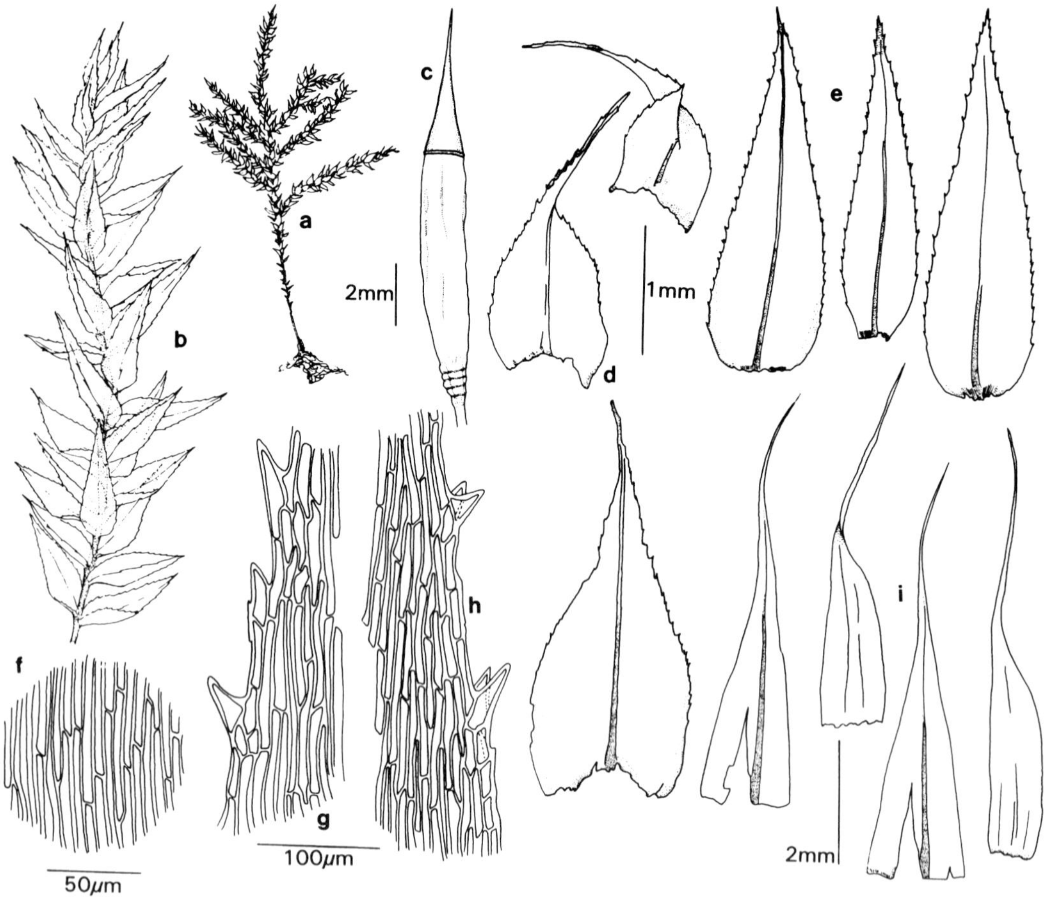

Fig. 9. *Hypnodendron reinwardtii* (Schwaegr.) Jaeg. (a, b from *Norris 64190*, UC; c from *Koponen 35566*, H; d–i from *Streimann 19664*, H). — a: Growth habit. — b: Branch. — c: Capsule. — d: Stipe leaves. — e: Branch leaves. — f: Leaf cells. — g: Leaf border of stipe leaf. — h: Leaf border of branch leaf. — i: Perichaetial leaves. — Use the 2 mm scale left up for c, 1 mm scale for d and e, the 50 μm scale for f, the 100 μm scale for g and h and the 2 mm scale right down for i.

18821 (H). *10. 11.* Present report. *12.* Present report, and: Spreader Divide, 1 950 m on twigs, *Streimann 26045* (H). *23.* Solomon Islands. *29.* Kolombangara, top of Mt. Veve, 1 600–1 760 m, on branch in elfin forest, *Norris 49518* (H).

Total range: As 4: Ind Ma Phi PNG; Oc: Sam.

10. *Hypnodendron subspininervium* (C. Müll.) Jaeg. (Figs. 10, 12)

in Jaeg. & Sauerb., Ber. Thät. St. Gall. Naturw. Ges. 1877/78: 359. 1879 (Adumbratio 2: 623). — *Hypnum subspininervium* C. Müll. — Bot. Zeit. 15: 782. 1857.

Trachyloma arborescens Mitt., J. Proc. Linn. Soc. Suppl. Bot. 1: 91. 1859. — *Hypnodendron arborescens* (Mitt.) Lindb. in Seemann, Flora Vitiensis 401. 1873. — *H. subspininervium* subsp. *arborescens* (Mitt.) Touw, Blumea 19: 237. 1971. — Reduced to subspecific status under *H. subspininervium* by Touw (1971).

Hypnodendron pseudo-arborescens Fleisch., Musci Fl. Buitenz. 1607. 1923. — Synonymized by Touw (1971).

Plants pale green to yellow green, prostrate to laxly erect, with a poorly defined stipe with rhizoids restricted to the base, and with spreading to reflexed leaves, to 10 cm long, pinnately to weakly bipinnately branched with conspicuous reddish stems and with branches to 1.5 cm long. Upper stipe leaves grading imperceptibly in form and size into the leaves of the larger branches, distant to rather closely inserted,

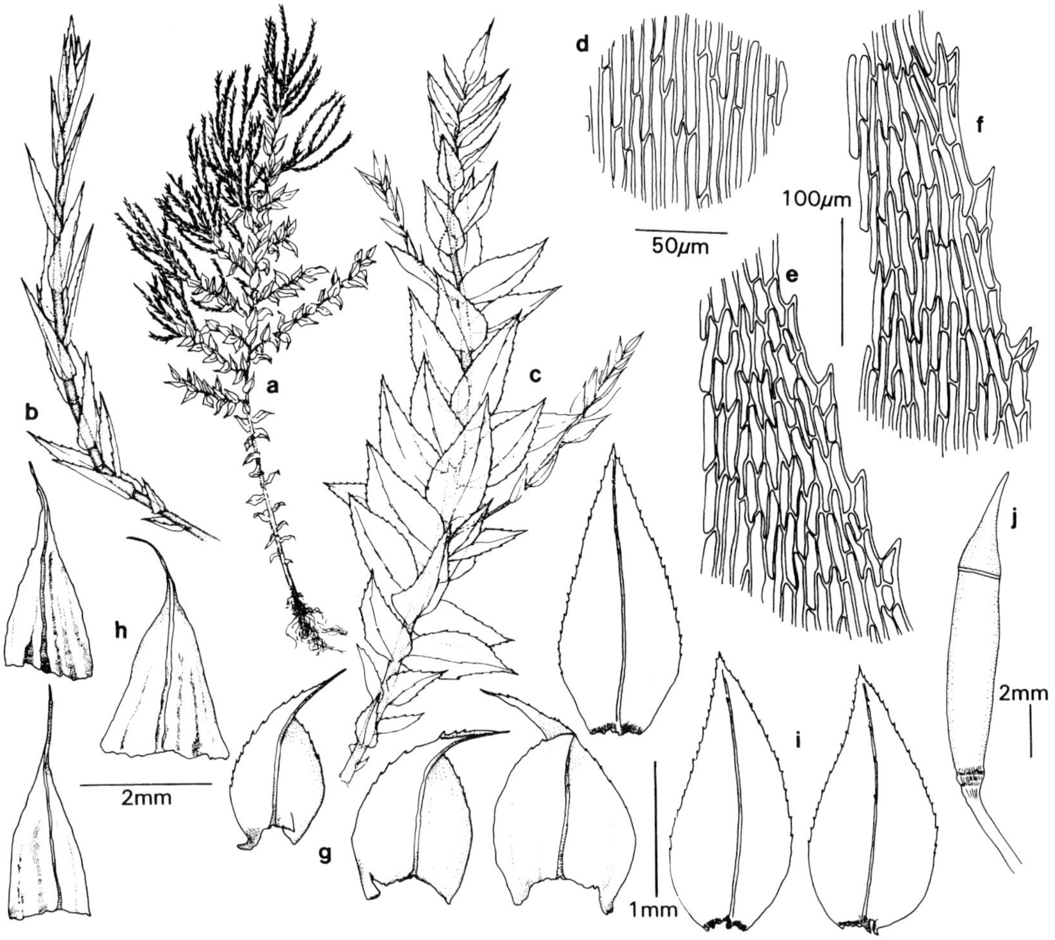

Fig. 10. *Hypnodendron subspininervium* (C. Müll.) Jaeg. (a–g, i from *Norris 66205*, H; h from *Norris 66081*, H; j from *Touw 21223*, H). — a: Growth habit. — b: Caducous branch. — c: Branch. — d: Leaf cells. — e: Leaf border of stipe leaf. — f: Leaf border of branch leaf. — g: Stipe leaves. — h: Perichaetial leaves. — i: Branch leaves. — j: Capsule. — Use the 50 µm scale for d, the 100 µm scale for e and f, the 2 mm scale on the left for h, the 1 mm scale for g and i and the 2 mm scale on the right for j.

broadly ovate, gradually narrowed to an acumen which occupies about 1/4–1/6 of leaf, to 2.5 mm long, 1.6–2:1, asymmetric, so keeled basally as to resist lying flat on a slide. Ultimate branch leaves approximate to imbricated, loosely spreading, not strongly keeled, to 1.8 mm long, about 2.5:1, broadest near the base and less strongly acuminate than the stipe leaves. Median laminal cells to 5 µm broad, 6–15:1, smooth to prorate, rather thin-walled. Alar cells not clearly differentiated from the other basal cells, these often markedly enlarged across the leaf, to 20 µm broad, 1.5–3:1, rather thick-walled and pitted, often red brown. Costa subpercurrent, usually with a few dorsal teeth. Margins basally slightly recurved, dentate with some teeth geminate throughout the distal 1/2–3/4 (more strongly so in the branch leaves) with teeth mostly unicellular. Axillary hairs to 150 µm, not offset from leaf insertion, 3–4 cells long with 2 basal brown cells. Stems rounded pentagonal in section, with a strong central strand surrounded by moderately thick walled interior cortical cells to 25 µm in diameter; these grading into several layers of highly pachydermous much smaller (to 8 µm) outer corticals. Pseudoparaphyllia linear, to 80 μm long, to 10 cells long and 3 cells wide at base, often overarching the macronematal apparati. Macronematal apparatus elliptical, to 150 μm wide, about 2:1, present at leaf axils outside the imaginary lines of decurrency from some of the stipe leaves.

Dioicous. Seta 3–5 cm long, brown to reddish brown, mostly at least somewhat flexed distally. Seta cross-section round with about 5 layers of clear white pachydermous cells inwardly defined by a ring of more brownish cells and then several layers of only slightly larger leptodermous cells and an indistinct central strand. Urn to 6 mm, cylindric, 4–6:1, mostly somewhat arcuate, usually lightly sulcate at least in distal portions. Operculum rostrate, to 1/2 of urn length. Exothecial cells to 25 μm wide, elongate hexagonal to rectangular, 1.5–3:1, with walls to 4 μm thick evenly thickened throughout with no corner thickenings. Sporangial wall to 125 μm thick with 5 layers of cells without a striping of large and smaller cells. Stomata superficial, restricted to short neck of capsule. Annulus vesiculose, not revoluble. Exostome to 800 μm long, dorsal plate more narrow than the ventral one, horizontally closely striate, with a zig-zag median commisural line. Endostome sparsely papillose, nearly as long as the exostome, with basal membrane comprising 1/2–2/3 of its length, processes keeled and perforate, cilia 2–5. Spores smooth or nearly so, to 25 μm.

Illustrations: Dozy & Molkenboer 1865: tab. CCXXXII; Touw 1971: 234 (fig. 4).

Hypnodendron subspininervium is not always distinguishable with confidence from *H. reinwardtii*. One should examine the length/width ratio of upper stipe leaves on lowland plants with a pinnate growth form, and with divergent and dentate leaves on a stipe mostly naked of rhizoids. In comparison with *H. reinwardtii*, this is a broader leaved plant and it seems to be restricted to the lowlands. In those lowlands, it may be easily distinguished from most *H. reinwardtii*, but highland examples of the latter species may be almost as broad leaved as *H. subspininervium*.

The teeth of the branch leaves of *Hypnodendron subspininervium* are usually somewhat shorter and more blunt than those of *H. brevipes* and *H. reinwardtii*. The alar cells may be a little less thick-walled and pitted. Median leaf cells of *H. subspinervium* are usually somewhat shorter than those of *H. reinwardtii*, and the walls may be a little thinner.

All of our specimens of *Hypnodendron subspininervium* were found at 750–1 100 m in primeval rainforest. The substrates of our 7 specimens are: tree and twigs (5) and log (2).

Range on the Huon Peninsula: *5c*. 66050, 66052, 66081, 66122, 66168, 66205. *14e*. 27902.

Range in Western Melanesia: West Irian. *1*, *7*, *8*. Papua New Guinea. *9*. Busilmin, 1 450–1 550 m, on tree trunks, *Touw 14894* (H). *10*, *11*. *12*. Present report, and: Mt. Kaindi Rd., 1 450 m, on rock, *Streimann 33326* (H). *14*. *19*. Hepataewa, 1 200 m, on tree trunk, *Streimann 33877* (H). *20*. K.B. Sawmill, 600 m, on vine, *Streimann & Naoni 16418* (H). *23*.

Total range: As 3: In Sri Tha; As 4: Ind Ma Phi PNG; Oc: Fij NC Sam Sol Van.

11. *Hypnodendron vitiense* Mitt. (Figs. 11, 12)

in Seemann, Flora Vitiensis 401. 1873 ("*Vitiense*").

Hypnodendron angustirete Dix., Ann. Bryol. 10: 16. 1937. — Type: Papua New Guinea. Morobe, Finisterre Range, 800–1 000 m, *Eiffert 41* (BM, holotype, not seen). — Synonymized by Touw (1971).

Hypnodendron ambiguum Broth. in Schum. & Lauterb., Nachtr. Fl. Deutsch. Schutzgeb. Südsee 34. 1905, nom. nud. — Original collection: Papua New Guinea. Morobe, Sattelberg, VII.1890 *Weinland* (not seen). — Synonymized by Touw (1971).

Plants glossy, green to pale green, erect to ascending with a well-defined stipe with rhizoids restricted to the base, to 8 cm long, pinnately to palmately branched with branches to 2.5 cm long, never with caudate or microphyllous branch tips. Leaves ascending and appressed in lower part of stipe but with at least the distal portions of the upper stipe leaves somewhat spreading, imperceptibly grading in form and size to the leaves of the larger branches. Stipe leaves distant, deltoid-ovate, acuminate above or with the acumen rather poorly defined, to 2.5 mm long, 1.6–2:1, somewhat asymmetric, little keeled. Ultimate branch leaves complanate, often with the ventral leaves smaller and more narrow than the dorsal leaves, loosely spreading, not strongly keeled, to 1.8 mm long, about 2.5:1, broadest near the base and less strongly acuminate than the stipe leaves. Median laminal cells to 5 μm broad, mostly more than 10: 1, smooth to rather low prorate, rather thin-walled. Alar cells not differentiated from adjacent basal cells, to 20 μm broad, 2–4:1, somewhat thicker walled and more pitted than median laminal cells. Costa subpercurrent, usually with a few dorsal teeth.

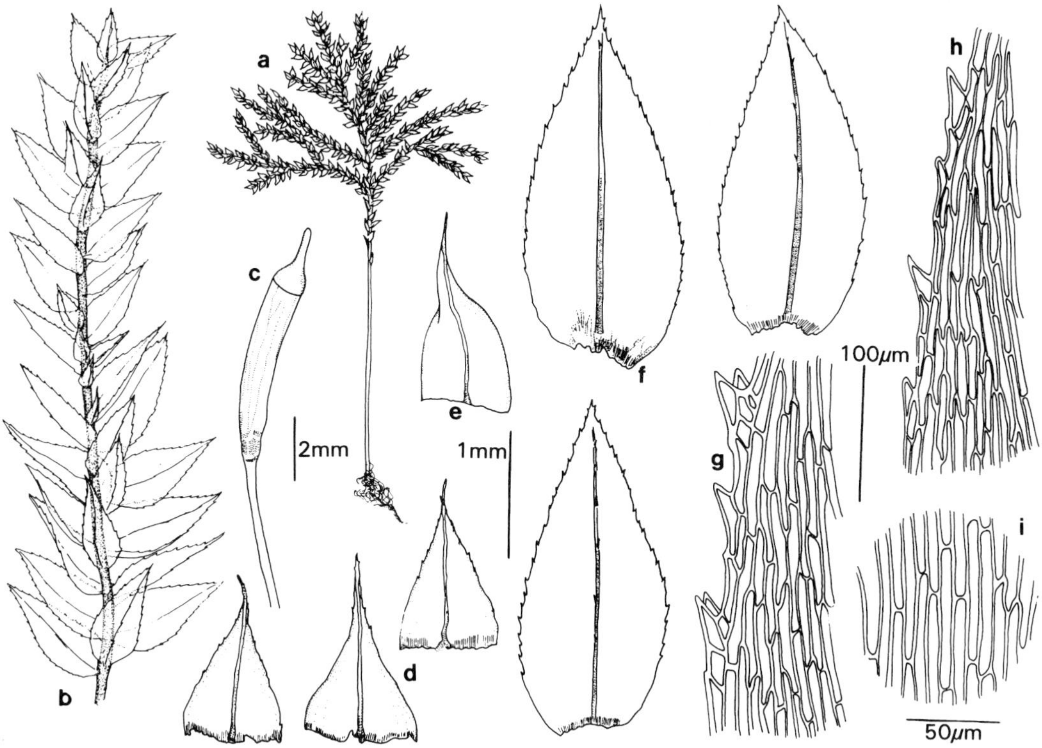

Fig. 11. *Hypnodendron vitiense* Mitt. (a, b, d, f–i from *Norris 59041*, H; c, e from *Streimann 056221*, H). — a: Growth habit. — b: Branch. — c: Capsule. — d: Stipe leaves. — e: Perichaetial leaf. — f: Branch leaves. — g: Leaf border of branch leaf. — h: Leaf border of stipe leaf. — i: Leaf cells. — Use the 2 mm scale for c, the 1 mm scale for d, e and f, the 100 µm scale for g and h and the 50 µm scale for i.

Margins of stipe leaves entire below and serrulate above with teeth mostly formed only of cell angles; margins of branch leaves with low, often geminate, teeth mostly consisting of only a single cell. Margin plane, not decurrent. Axillary hairs to 80 µm long, only slightly offset from leaf insertion, with 2 basal brown cells and two longer terminal cells to 30 µm long. Stems rounded pentagonal in cross-section, with a strong central strand, surrounded by leptodermous and large (to 25 µm) inner cortical cells which grade into the very pachydermous red and small (to 8 µm) outer corticals. Pseudoparaphyllia lanceolate and entire, to 50 µm long, 3–4 cells long and 2 cells wide at base, often overarching the macronematal apparati. Macronematal apparatus not axillary, present outside the decurrency on most of the stipe leaf, elliptic, to 80 µm wide, about 2:1, with short rectangular pale cells.

Dioicous. Seta 2–5 cm long, brown to reddish brown, mostly at least somewhat flexed distally. Seta cross-section elliptic, to 0.2 mm thick, of 2–3 layers of pachydermous cells and 5–6 layers of leptodermous larger inner cells and with the central strand very indistinct. Urn to 4 mm, cylindric, 3–5:1, arcuate, sulcate at least in distal portions. Operculum rostrate, to 1/3 of length of urn. Sporangial wall only 3 cells (75 µm) thick with the cells very irregular in size and shape. Exothecial cells quadrate to short rectangular, to 25 µm wide, 2–5:1; but with vertical streaks of narrower cells only 10 µm wide. Exostome to 700 µm long, dorsal plate more narrow than the ventral one, horizontally closely striate, with a zigzag median commisural line. Endostome nearly smooth, nearly as long as the exostome, with basal membrane comprising 1/2–2/3 of its length, processes keeled and perforate, cilia 2–4. Spores smooth or nearly so, to 15 µm.

Illustrations: Bartram 1939: pl. 12 (fig. 190); Touw 1971: 280 (fig. 23.)

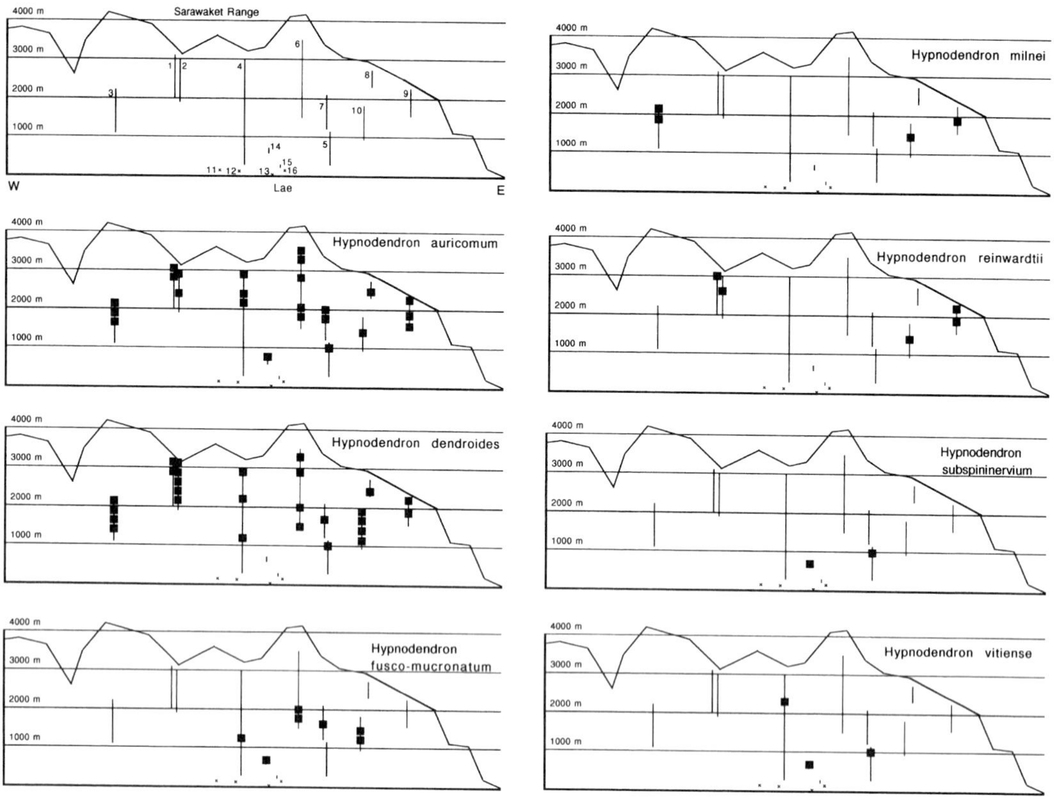

Fig. 12. The altitudinal ranges of *Hypnodendron auricomum* Broth. & Geh., *H. dendroides* (Brid.) Touw, *H. fuscomucronatum* (C. Müll.) Jaeg., *H. milnei* Mitt., *H. reinwardtii* (Schwaegr.) Jaeg., *H. subspininervium* (C. Müll.) Jaeg., and *H. vitiense* Mitt. on the Huon Peninsula. Compare with fig. 4 in Koponen and Norris (1983). Nos. 1–16 indicate collecting localities and altitudinal range of the Koponen and Norris expedition.

The configuration of the stipe leaves of *Hypnodendron vitiense* is rather difficult to describe. Like essentially all other works, we have keyed this species as having appressed leaves. Touw (1971) speaks of of them as "appressed to obliquely spreading", and this is accurate if we note that the leaves near the base are strongly appressed but those near the base of the branches may be so divergent as to tempt one to key through as "Leaves obviously spreading from the stem".

On the stipe leaves, the virtual lack of teeth composed of more than simple cell angles allows confusion of this plant only with poorly developed forms of *Hypnodendron menziesii*. The latter species may be distinguished because it has reflexed stipe leaves with a twisted acumen formed almost exclusively of the excurrent costa.

This is the only one of our New Guinea *Hypnodendron* which is also found in Australia. Similarly, it is unique in extending northward to the Ryukyus and to Taiwan.

Our four specimens of *Hypnodendron vitiense* were found in primeval forest at 750–2 400 m. The substrates are: rock or boulder (2) and log (2).

Range on the Huon Peninsula: *4g*. 66619. *5c*. 66156. *14e*. 58973, 59041.

Range in Western Melanesia: West Irian. *4*. Papua New Guinea. *10, 11. 12*. Present report. *14, 25, 27*. Solomon Islands. *29*. Kolombangara: Mt. Veve, 1 000–1 760 m, on trees and soil, *Norris & Roberts 49314, 49379, 49415, 49444, 49559, 49744, 49824* (H). *31. 33*. South of Manighai, 600–720 m, *Norris & Roberts 49044* (H).

Total range: As 2: Chi Ja; As 3: Vi; As 4: Ind Phi PNG; Oc: Fij NC Sam Sol Van. Austr 1.

Doubtful records and excluded taxa

Hypnodendron subarborescens C. Müll.

in Geh., Biblioth. Bot. 13: 10. 1889. — Type: Papua New

Guinea. Western, Fly River (Branch), 1885 *Bäuerlen 354* (not seen).

Hypnodendron arborescens (Mitt.) Lindb. var. *minus* Thér., Diagn. Esp. Var. Nouv. Mouss. 8: 7. 1910.

Range in Western Melanesia. West Irian. *2*. (Geheeb 1898). Papua New Guinea. *18*. (Geheeb 1889).

Touw (1971) had not seen the type of *Hypnodendron subarborescens*, and he considered the name of doubtful referent. We have not been able to clarify the problem.

Mniodendron nanum C. Müll.

Hedwigia 41: 133. 1902. — *M. nanum* C. Müll. ex Par., Ind. Bryol. 250. 1900, nom. nud. — *Hypnodendron nanum* C. Müll. ex Par., Ind. Bryol. 192. 1900, nom. nud. — Type: Papua New Guinea. Central "Nova Guinea austro-orientalis, Mo-roka, 1 100 m, Aug. 1893: Lamberto Loria. Hb. Levier" (not seen).

Brotherus (1909) proposed that *Mniodendron nanum* might not be specifically different from *Mniodenron parvum* C. Müll. (= *Hypnodendron milnei*), and Dixon (1922) repeated Brotherus' opinion. Judging from the description *Mniodendron nanum* is near to or identical with *Hypnodendron fuscomucronatum* (Touw 1971). Touw saw 5 Loria specimens named by C. Müller, but none of them contain *H. fuscomucronatum*. We have not been able to clarify the problem.

Acknowledgements. We especially wish to thank Dr. Pekka Isoviita for valuable advice on nomenclature and bibliography. We also thank the Curators of the Herbaria from which we received specimens on loan (BM, CBG, FH, L, LAE). Mrs Inkeri Ahola, MSc, (Helsinki) and Juliana Duncan (Arcata) are thanked for drawing of the figures. Grants from the Academy of Finland and from the Finland–United States Educational Exchange Commission are also acknowledged.

REFERENCES

Akiyama, H. 1989: A reevaluation of Hypnodendron opacum (Hypnodendraceae, Musci). — Bryologist 92: 198–201.

Bartram, E. B. 1939: Mosses of the Philippines. — Philippine J. Sci. 68: 1–147.

Bartram, E. B. 1965: Mosses of the Eastern Highlands, New Guinea, from the 6th Archbold Expedition, 1959. — Contrib. U. S. Natnl. Herb. 37: 43–67.

Brotherus, V. F. 1898: Some new species of Australian mosses described IV. — Öfvers. Finska Vetensk. Soc. Förhandl. 40: 159–193.

Brotherus, V. F. 1909: Unterklasse Bryales. II. Spezieller Teil. — In: Engler, A. & Prantl, K. (eds.), Die natürlichen Pflanzenfamilien 1(3): 277–700. W. Engelmann, Leipzig. 700 pp.

Brotherus, V. F. 1924, 1925: Musci. — In: Engler, A. & Prantl, K. (eds.), Die natürlichen Pflanzenfamilien, ed. 2, 10: 129–131, 143–478 (1924); 11: 1–522 (1925). W. Engelmann, Leipzig. 10: 478 pp. 11: 542 pp.

Buck, W. R. & Vitt, D. H. 1986: Suggestions for a new familial classification of pleurocarpous mosses. — Taxon 35: 21–60.

Dixon, H. N. 1922: The mosses of the Wollaston Expedition to Dutch New Guinea, 1912–13; with some additional mosses from British New Guinea. — J. Linn. Soc. Bot. 45: 477–510.

Dozy, F. & Molkenboer, J. M. (fortgesetzt von R. B. van den Bosch & C. M. van der Sande Lacoste) 1855–70: Bryologia Javanica seu descriptio muscorum frondosum Archipelagi Indici. 1–2. — E. J. Brill, Lugduni-Batavorum (Leiden). 1: 161 pp., pls. 1–130; 2: 238 pp., pls. 131–320.

Fleischer, M. 1917: Die Laubmoose Papuasiens. I. — Bot. Jahrb. 55: 19–37.

Fleischer, M. 1922: Die Musci der Flora von Buitenzorg (zugleich Laubmoosflora von Java). 4: I–XXXI, 1105–1729. — E. J. Brill, Leiden.

Geheeb, A. 1889: Neue Beiträge zur Moosflora von Neu-Guinea. — Biblioth. Bot. 13: 1–13.

Geheeb, A. 1898: Weitere Beiträge zur Moosflora von Neu-Guinea. — Biblioth. Bot. 44: 1–25.

Koponen, T. 1990: Bryophyte flora of Western Melanesia. — Trop. Bryol. 2: 149–160.

Koponen, T. 1993: The Project "Bryophyte flora of the Huon Peninsula, Papua New Guinea". — Bryol. Times 76: 6–8.

Koponen, T. & Norris, D. H. 1983: Bryophyte flora of the Huon Peninsula, Papua New Guinea. I. Study area and its bryological exploration. — Ann. Bot. Fennici 20: 15–29.

Koponen, T., Norris, D. H. & Piippo, S. 1992: Bryophyte flora of Western Melanesia. A status report. — Bryobrothera 1: 157–160.

Koponen, T., Piippo, S. & Norris, D.H. 1991: Bryophyte flora of the Huon Peninsula, Papua New Guinea. XLVII. Generic and family indexes to parts I–XLVI. — Acta Bot. Fennica 143: 107–111.

Lindberg, S.O. 1862 ('1861'): Om en ny art af slägtet Hypnum. — Öfvers. Kungl. Svenska Vet.-Akad. Förhandl. 18: 371–375.

Noguchi, A. 1953: Mosses of Mt. Sarawaket, New Guinea. — J. Hattori Bot. Lab. 10:1–23.

Norris, D. H. & Koponen, T. 1985a: Bryophyte flora of the Huon Peninsula, Papua New Guinea. VII. Trachypodaceae, Thuidiaceae and Meteoriaceae (Musci). — Acta Bot. Fennica 131: 1–51.

Norris, D. H. & Koponen, T. 1990: Bryophyte flora of the Huon Peninsula, Papua New Guinea. XXXIII. Leskeaceae and Fabroniaceae (Musci) plus corrigenda and addenda to previous papers. — Ann. Bot. Fennici 27: 1–12.

Piippo, S., Koponen, T., Hyvönen, J. & Enroth, J. 1994: Länsi-Melanesian sammalten monimuotoisuuden tutkimus. — Luonnon Tutkija 98: 198–203.

Sainsbury, F. L. S. 1955: A handbook of the New Zealand mosses. — Bull. Roy. Soc. New Zealand 5: 1–490.
Touw, A. 1971: A taxonomic revision of the Hypnodendraceae (Musci). — Blumea 19: 211–354.
Zanten, B. O. van 1964: Mosses of the Star Mountains Expedition. — Nova Guinea Bot. 16: 263–368, pls. XXII–XXXVI.

INDEX TO NAMES

An asterisk means that the lectotype has been proposed in this paper. The accepted names of taxa occurring in Western Melanesia are in **boldface**.

Acta Bot. Fennica 156: 49–62 ISBN 951-9469-50-8 ISSN 0001-5369
Helsinki 4 September 1996

Bryophyte flora of the Huon Peninsula, Papua New Guinea. LIX. *Breutelia* (Bartramiaceae, Musci)

Viivi Virtanen

Virtanen, V., Department of Ecology and Systematics, P.O. Box 47, FIN-00014 University of Helsinki, Finland

Received 17 November 1995

The genus *Breutelia* (B. S. G.) Schimp. includes 6 species in Western Melanesia: *B. arundinifolia* (Duby) Fleisch., *B. crassicaulis* (C. Müll.) Par., *B. aristifolia* Zanten, *B. longicapsularis* Dix., *B. roemeri* Fleisch., and *B. papuensis* Virtanen, which is described as a new species. All except *B. arundinifolia* and *B. crassicaulis* are endemic to New Guinea. A key to Western Melanesian species is given, with species descriptions and illustrations. The taxonomy, ecology and distribution of the species are discussed.

Key words: Bartramiaceae, *Breutelia*, Musci, nomenclature, Papua New Guinea, Solomon Islands, taxonomy, West Irian

INTRODUCTION

This paper belongs to a series dealing with the bryophyte flora of Western Melanesia, which includes West Irian, Papua New Guinea and Solomon Islands. Essential background information of material and methods, and abbreviations of collecting localities and geographical areas used in this study, are given in part I (Koponen & Norris 1983) and VII (Norris & Koponen 1985). This study is mainly based on the collections of Koponen and Norris from the Huon Peninsula, Papua New Guinea (H). Material from other herbaria has also been studied, and the specimens are cited in the text. I have studied the type specimens of the species recorded for Asia and Oceania except *Breutelia crassicaulis* (C. Müll.) Par. and *B. dicranacea* (C. Müll.) Mitt. For the range in Western Melanesia only specimens for new records are cited, and the records indicated by numbers *1–33* are based on an unpublished manuscript by S. Piippo, T. Koponen and D. H. Norris (Annotated catalogue of Western Melanesian bryophytes. II. Musci). Genera and family indices to parts I–XLVI are given in Koponen *et al.* (1991). Scanning electron microscope (SEM) was used to study the surface structures of leaves, peristomes and spores of three species: *B. arundinifolia* (Duby) Fleisch., *B. papuensis* Virtanen and *B. roemeri* Fleisch.

FAMILY BARTRAMIACEAE SCHWAEGR. 1830

According to Griffin and Buck (1989) the Bartramiaceae consists of 11 genera, *Anacolia* Schimp., *Bartramia* Hedw., *Breutelia* (B. S. G.) Schimp., *Catoscopium* Brid., *Conostomum* Sw., *Fleischerobryum* Loesk., *Flowersia* Griffin & Buck, *Leiomela* Mitt., *Philonotis* Brid., *Plagiopus* Brid., and *Quathlamba* Magill, most of the species belonging to the genera

Bartramia, *Breutelia*, and *Philonotis*. Brotherus (1904) included 9 genera in the family. Modern revision of the family is lacking but regional revisions have been made of East Asian (Kabiersch 1937) and Malaysian (Manuel 1981) genera. The spore morphology was studied by Hirohama (1977), surface ornamentation of rhizoids by Hirohama and Iwatsuki (1980), stem anatomy by Kawai (1982), axillary hair morphology by Griffin and Buck (1989), and branch development by Akiyama and Nishimura (1993).

Genus **Breutelia** (B. S. G.) Schimp. 1856

A modern taxonomic revision of the genus *Breutelia* of Bartramiaceae has not been made, and the number of accepted species based on Index Muscorum (Wijk *et al.* 1959, Crosby *et al.* 1992, and Crosby & Magill 1994) is 125. Sloover (1975) and O'Shea (1995) reported 20 species and one variety from sub-Saharan Africa and adjacent islands. Griffin (1984) reported ca. 65 species in Central and South America, including 7 species from Patagonia studied by Matteri (1973). In Asia (As 2, 3, 4) and Oceania, 15 species have been recognized (Wijk *et al.* 1959, Crosby *et al.* 1992, Crosby & Magill 1994).

The genus is defined by subperichaetial innovations, plicate leaves, differentiated alar regions with usually isodiametric or rectangular cells, a single, often excurrent, strong costa, linear laminal cells, lanceolate-linear leaves, and mostly horizontal to pendulous and cylindrical capsules. All species are dioicous. However, one specimen of *Breutelia roemeri* had both a sporophyte and antheridia. The perigonia are terminal and discoid. The peristome is diplolepideous and of the bryaceous type with 16 lanceolate exostome teeth positioned alternate to 16 endostome segments arising from a high basal membrane. Each segment is separated by cilia, which may sometimes be lacking. Operculum is conical. The species present on the Huon Peninsula are fairly large and easy to identify. In New Guinea all six species belong to montane floristic element growing mostly epigaeic in open tussock grasslands, or in more shady mossy forests at altitudes from 1 500 m to 3 600 m.

Key to the species of *Breutelia* in Western Melanesia

1. Leaves with subclasping base, spreading or recurved .. 2
— Leaves without subclasping bases, erect to erecto-patent .... 5. *B. crassicaulis*
2. Leaves ovate-lanceolate, aristate .... 1. *B. aristifolia*
— Leaves linear-lanceolate, not aristate 3
3. Leaves linear, > 6.0 mm, the whole subclasping base intensive red .... 2. *B. longicapsularis*
— Leaves linear-lanceolate, < 6.0 mm, only insertion mostly reddish 4
4. Leaves > 5.2 mm, mostly 5.7 mm, alar cells 37–60 µm long and 25–38 µm wide .... 3. *B. arundinifolia*
— Leaves < 5.2 mm, mostly 3.6–3.9 mm, alar cells 30–50 µm long and 20–25 µm wide 5
5. Length of apical cells > 10 µm .... 4. *B. roemeri*
— Length of apical cells < 10 µm .... 6. *B. papuensis*

1. *Breutelia aristifolia* Zanten (Figs. 1, 7, 9)

Nova Guinea Bot. 16: 294. *Pl. 24 (fig. 4).* 1964. — Type: Indonesia. West Irian, Star Mts., Mt. Antares, west summit, 3 400 m, *Zanten 676* (L!, holotype).

Plants fairly robust, 3–10 cm high, bright yellow to golden brown; stems in loose tufts or nearly solitary, creeping and ascending, irregularely branched, branches usually few, sometimes heavily tomentose below; apices blunt, usually yellow, lighter than other parts of the plants. Leaves crowded, clasping, squarrose and tightly appressed with tips pointing downwards when dry, squarrose-spreading when moist, 3.2–4 mm long, ca. 1.8 mm at widest point, ovate-lanceolate, abruptly narrowed to long hyaline aristate apex, base shortly and deeply plicate, with red or brown colour in lower parts. Margin bluntly serrulate from the apex to the mid-leaf. Costa 3–4 cells broad, 0.25–0.5 mm wide basally, very conspicuously excurrent but often broken, yellow to brown, not red. Leaf cells linear to vermiculate, distinctly papillose to smooth, single papillae present at the basal part of the cells, cell walls usually firm or slightly papillose. Apical cells 22–32 µm, median cells 25–45 µm, and basal cells 35–75 µm long and 2.5–5 µm wide. Alar cells hyaline, rectangular, 50–100 µm long and 25–45 µm wide, in 3–4 rows each having 6–10 cells.

Dioicous. Perigonial leaves broadly ovate to triangular, ca. 5.2 mm long and 3.2 mm at widest point, costa excurrent with ca. 140 µm long apex, margins serrate. Perichaetial leaves lanceolate, ca. 3.9 mm long and 1.4 mm wide basally, costa red in basal part and pale above, margins entire. Seta 2.5–3.5 cm long, red, slightly paler towards capsule. Capsules ovoid, horizontal to pendulous, 5–6 mm long and 2–3 mm wide, longitudinally furrowed or sulcate.

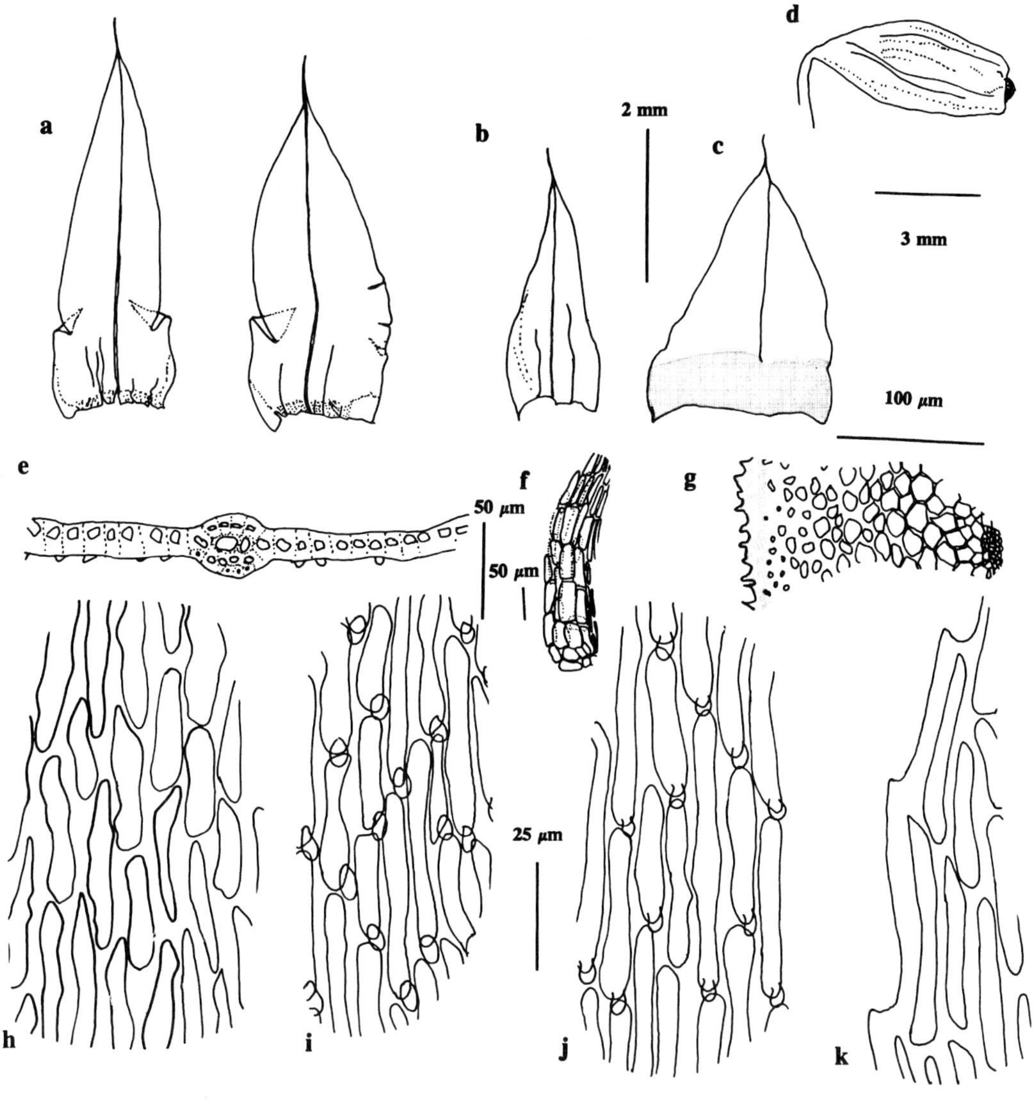

Fig. 1. *Breutelia aristifolia* Zanten (the stem leaf on right, c, h–k from *Norris 62292;* b and d from *Veldkamp & Stevens 5845*, L; the stem leaf on left, e, f, g from *Koponen 32850*). — a: Stem leaves. — b: Perichaetial leaf. — c: Perigonial leaf. — d: Capsule. — e: Leaf cross-section. — f: Alar cells. — g: Stem cross-section. — h: Apical leaf cells. — i: Median leaf cells. — j: Basal leaf cells. — k: Apical leaf margin. — Use the 2 mm scale for a–c, the 3 mm scale for d, the 50 μm scale on the left for e and on the right for f, the 100 μm scale for g, and 25 μm scale for h–k.

Exostome teeth linear-lanceolate, ca. 0.5 mm long and 0.14 mm wide basally, endostome segments ca. 20 μm longer than exostome teeth, splitting, strongly papillose in upper half, many papillae in each cell column. Cilia were not seen in material but they may exist. Spores smooth, ca. 20 μm in diameter.

Illustrations: Zanten 1964: pl. 24 (fig. 4).

Breutelia aristifolia is unlikely to be mistaken for any other species of *Breutelia*. The tightly appressed, squarrose leaves with tips pointing downwards are distinct from those of the other species with the leaves more or less spreading. Also the long excurrent costa and ovate leaves are distinguishing features. Sometimes *B. roemeri* with tightly appressed and squarrose leaves can be confused with *B. aristifolia*, but clearly porose cell walls, linear leaves (compared with ovate leaves of *B. aristifolia)*

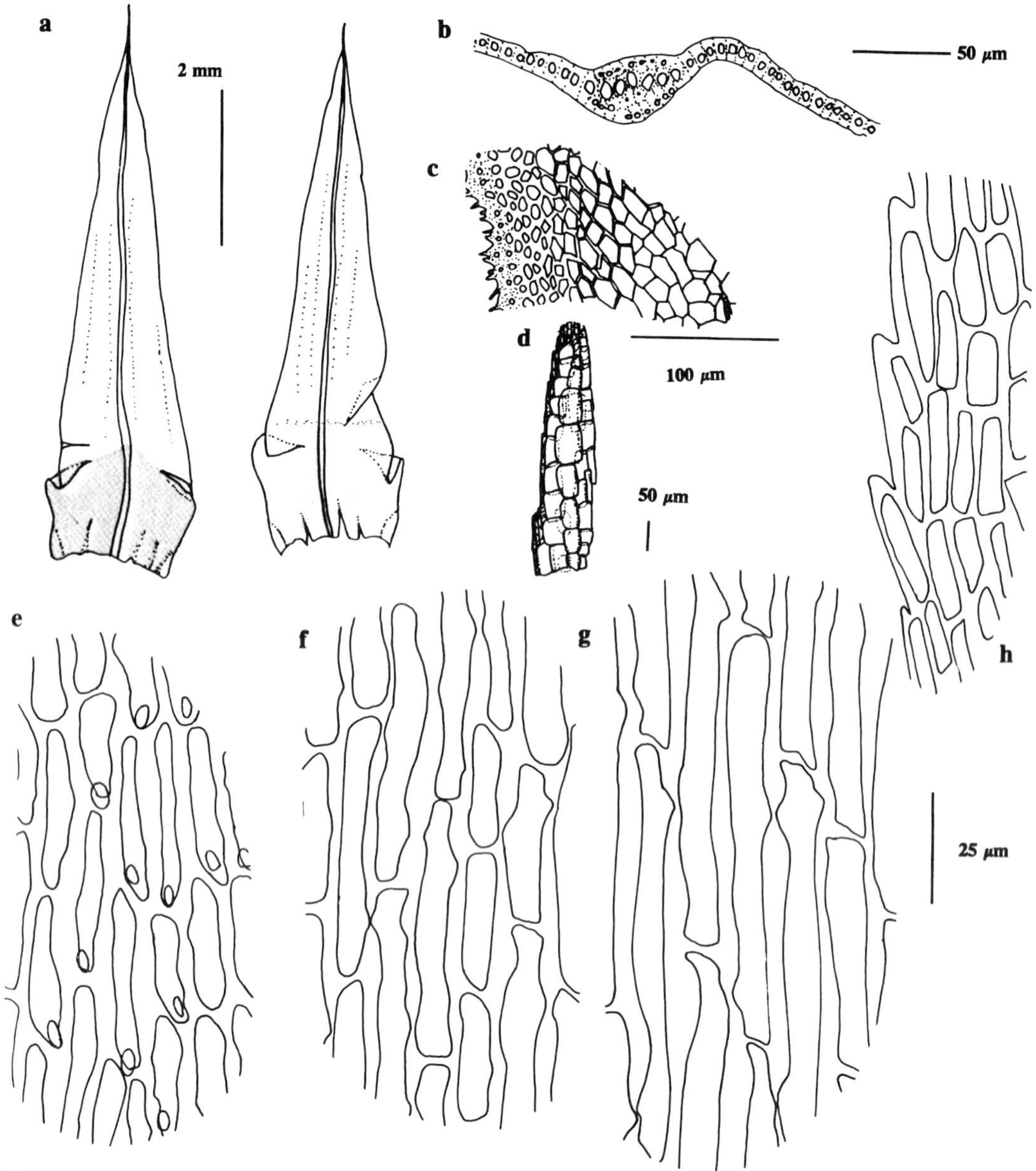

Fig. 2. *Breutelia longicapsularis* Dix. (from *Koponen 32559*. — a: Stem leaves. — b: Leaf cross-section. — c: Stem cross-section. — d: Alar cells. — e: Apical leaf cells. — f: Median leaf cells. — g: Basal leaf cells. — h: Apical leaf margin. — Use the 2 mm scale for a, the 50 µm scale on the left for b and on the right for c, the 100 µm scale for d, and the 25 µm scale for e–h.

and more sharply dentate leave margins are distinguishing features of *B. roemeri*. *Breutelia aristifolia* seems to be common in New Guinea. It is a species of open sites, and it occurs mainly in grasslands and bogs, and it was not found in mossy forest like the other species of the genus in New Guinea. It was found only epigaeic or at bases of shrubs. On the Huon Peninsula *B. aristifolia* was collected on alpine grassland with scattered treeferns at the altitudes of 2 850 m to 3 600 m in open, and mesic or moist sites. The substrates were soil (5 specimens), humus (2), peat (1) and dry stream bed (1).

Range on the Huon Peninsula: *6a*. 62292, 62400. *6b*. 31886. *6d*. 32166. *6e*. 62883. *6f*. 32336. *6j*. 63156. *6m*. 32850. *6u*. 62729B. *6x*. 63131.

Range in Western Melanesia: West Irian. *6*. Carstenz

Mts., on *Rapnaea* shrubs, 3 840 m, *Hope CGE M93* (L). *7.* Type (see above). Papua New Guinea. *10.* Mt. Burgers, on blanket bog, 3 675 m, *Veldkamp & Vinas 7491A* (L). *12.* Mt. Bangeta, in alpine vegetation, 4 100 m, 1963 *Royen* (H). *14.* Mt. Aas, alpine grassland, in wet places, 3 740 m, *Vink 16280A* (L). *15. 16.* Mt. Wilhelm, upper Pinde valley, on inundated swampy bog, 3 700 m, *Balgooy 792E* (L). *17.* Mt. Kerewa, on disturbed and burnt area on peat, *Veldkamp & Wiakabu 7776C* (L). *18.* Mt. Karoma, on *Styphelia* heath, *Veldkamp & Wiakabu 7737B* (L). *21.* Goropu Mts., Mt. Suckling, in damp places, *Veldkamp & Stevans 5845* (L).

Total range: Endemic to New Guinea.

2. *Breutelia longicapsularis* Dix. (Figs. 2, 7, 9)

J. Linn. Soc. Bot. 45: 482. 1922. — Type: Indonesia. West Irian, Paniai, Mt. Carstensz, XIII Camp, 10 300 ft., 29.I. 1913 *Boden Kloss* (BM!, holotype).

Plants very robust, 10 to 15 cm long, yellow or sometimes bright (light) green above and brown below. Stems solitary, or in loose tufts, distinctly red but hidden by red leave bases, unbranched or with only a few branches. Leaves 6.0–7.3 mm long, linear (especially above the red base), with four to six conspicious long striae starting above the base, the whole subclasping base is distinctly red to orange, and ca 1.3 mm long (1/5 of the leaf length), apices spreading. Leaf margin denticulate from the mid-leaf to the apex, by the apex teeth sharper and present also on the excurrent costa. Costa with 0.4 mm long excurrent apex, ca. 63 µm broad, consisting of 5–6 cell rows, red, or rarely yellow. Cells linear, sometimes more rectangular and clearly smaller in the apex, mostly distinctly papillose, papillae present at the basal part of the cells. Cell margins mostly smooth but sometimes slightly papillose, in basal part cells sinuose and cell walls distinctly incrassate, thicker than lumen. Apical cells 20–40 µm, median cells 30–60 µm, and basal cells 50–100 µm long and 2.5–5 µm wide. Alar cells hyaline, rectangular, much larger (wider) than laminal cells, 37.5–60 µm long and 25–37.5 µm wide forming a distinct group 4–5 cells wide and 6–8 cells long.

Dixon (1922) described the sporophyte, and the type specimen was the only specimen with sporophyte available for the study.

Breutelia longicapsularis is the largest species of the genus in the area. It is easily identified by the clearly linear leaves and the intensively colored basal part of leaves. Due to its large size *B. longicapsularis* could be confused with *B. arundinifolia*, but the latter species has linear-lanceolate leaves and only the very lowest part of the leaf base is red.

The most typical habitat of *Breutelia longicapsularis* is open tussock grassland, though it has been occasionally collected also in mossy forests. Wade and McVean (1969) reported the species from the altitudes of 3 350–4 200 m, in lower and upper subalpine forest and alpine fern meadow, which is associated with alpine grassland. In upper subalpine forest, 3 590–3 900 m, *B. longicapsularis* was common and they classified it as a constant species. On the Huon Peninsula *B. longicapsularis* occurred on open grassland with scattered treeferns at the altitudes of 2 850–3 550 m, growing on soil (2 specimens), humus (1) and cliff (1).

Range on Huon Peninsula: *6a.* 62358. *6b.* 62439. *6e.* 32290. *6j.* 32559 (with *B. aristifolia*). *6x.* 63041.

Range in Western Melanesia: West Irian. *6. 7.* Papua New Guinea. *9. 12.* Mt. Amungwiwa, in subalpine shrubberies, 3 220 m, *Royen 11159* (L). *13. 14. 15.*

Total range: Endemic to New Guinea.

3. *Breutelia arundinifolia* (Duby) Fleisch. (Figs. 3, 7, 9)

Musci Fl. Buitenzorg 2:630. *120.* 1904. — *Hypnum arundinifolium* Duby in Moritzi, System. Verzeichn. Zoll. Pfl.: 131. 1846. — *Bartramia arundinifolia* (Duby) C. Müll., Gen. Musc. Fr. 347. 1900. — Type: Indonesia. Java, monte Tjapul ca. 5 000 ft., 1844–1846 *Zollinger 1811* (G! holotype, H-BR! isotype).

Plants robust, 10 to 15 cm high, bright yellow to golden brown, apex usually yellow, lighter than other parts of the plants, stems in loose tufts, distinctly tomentose below, mostly creeping and ascending, simple or remotely dichotomously branched, number of branches variable. Leaves 5.2–5.7 mm long, linear-lanceolate, conspicuously deeply plicate, glossy, crowded, widely spreading from a subclasping base, stellate, often with reddish basal part. Margins recurved near base, margin entire below, serrate from the apex to the middle, teeth mostly blunt but sharper by the apex. Costa narrow, 3–4 cells broad, yellow, with ca. 340 µm long, excurrent apex, papillose. Cells linear, but apically almost rectangular, basal parts distinctly papillose to smooth. Apical cells 10–25 µm, median cells 30–50 µm, and basal cells 55–87 µm long, and 2.5–5.0 µm wide. Alar cells very big, hyaline, forming a group of 3–4 cell wide and 6–10 cell long, quadrate to rectangular, 37–60 µm long and 25–38 µm wide.

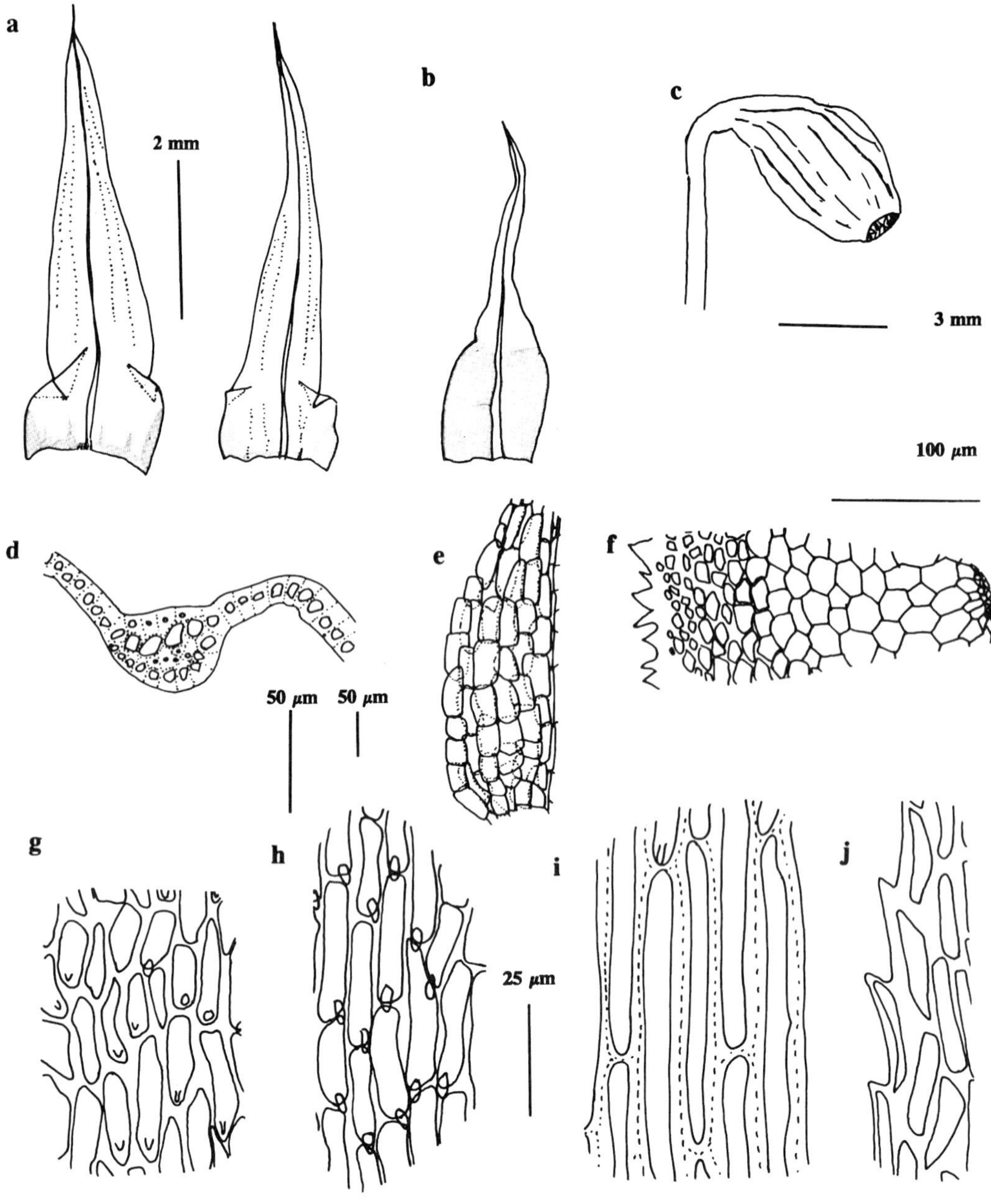

Fig. 3. *Breutelia arundinifolia* (Duby) Fleisch. (b and c from *Norris 65209*, others from *Koponen 34673*). — a: Stem leaves. — b: Perichaetial leaf. — c: Capsule (wet). — d: Leaf cross-section. — e: Alar cells. — f: Stem cross-section. — g: Apical leaf cells. — h: Median leaf cells. — i: Basal leaf cells. — j: Apical leaf margin. — Use the 2 mm scale for a–b, the 3 mm scale for c, the 50 µm scale on the left for d and on the right for e, the 100 µm scale for f, and the 25 µm scale for g–j.

Dioicous. Sporophytes rare. Perichaetial leaves linear-lanceolate, ca. 4.7 mm long and 1.3 mm wide basally, cells in basal half of the leave hyaline by the margins and reddish in the middle, apical cells light green and sinuate. Cell walls firm, in apical part cells papillose, costa with serrate, excurrent apex, leaf-margins entire below, and serrate above. Seta 3.5–5 cm long, red, sometimes dark reddish to black

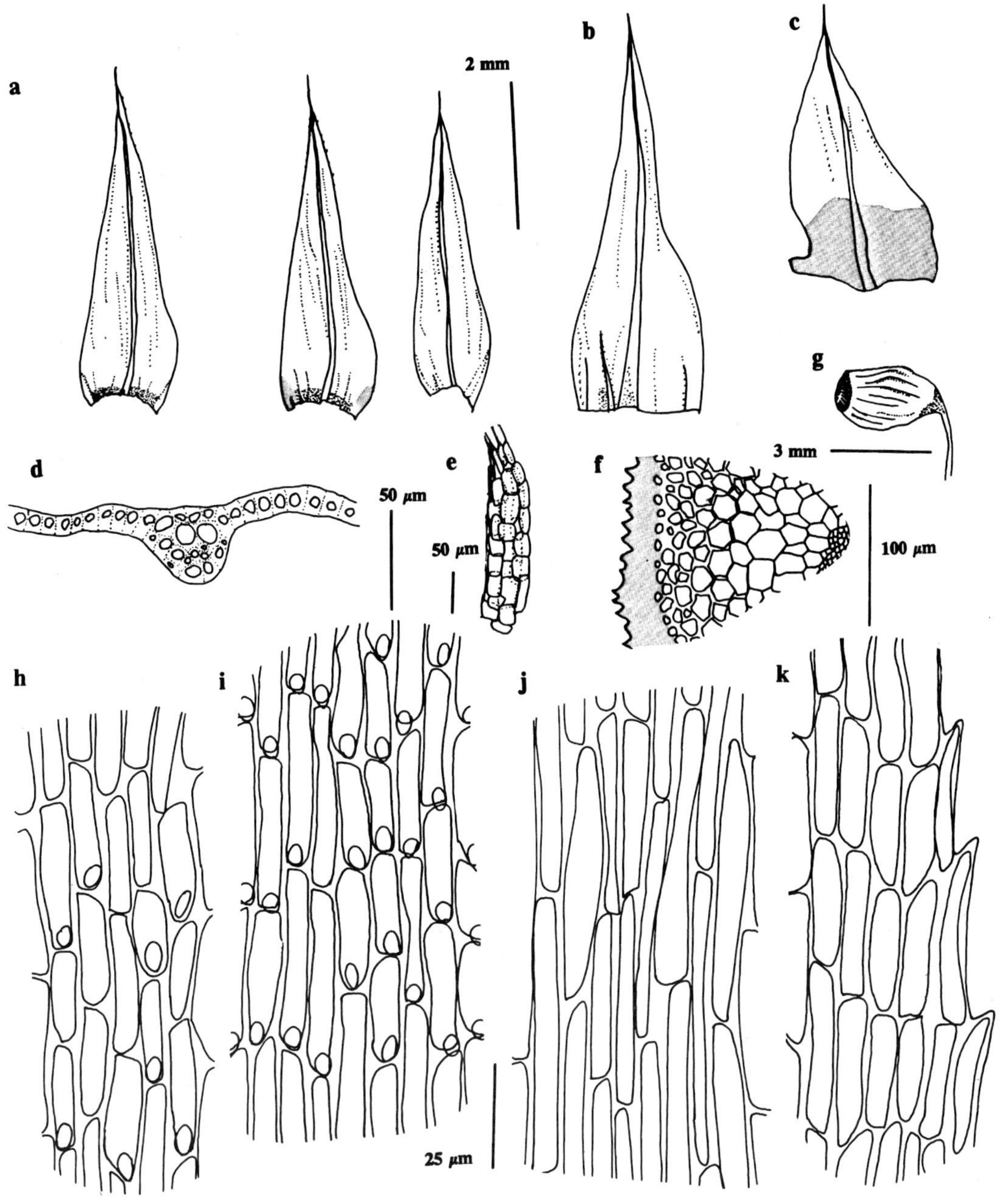

Fig. 4. *Breutelia roemeri* Fleisch. (b and d from *Koponen 31330;* c from *Norris 62815*; others from *Koponen 31312*). — a: Stem leaves. — b: Perichaetial leaf. — c: Perigonial leaf. — d: Leaf cross-section. — e: Alar cells. — f: Stem cross-section. — g: Capsule. — h: Apical leaf cells. — i: Median leaf cells. — j: Basal leaf cells. — k: Apical leaf margin. — Use the 2 mm scale for a–c, the 50 µm scale on the left for d and on the right for e, the 3 mm scale for g, the 100 µm scale for f, and 25 µm scale for h–k.

in top of seta and base of capsule, basally black. Capsule ovoid to cylindrical, horizontal, 5–7 mm long and ca. 2 mm wide, longitudinally distinctly furrowed. Exostome teeth linear-lanceolete, ca. 0.7 mm long and 0.19 mm wide basally, papillose in upper half, endostome segments as long as exostome teeth, splitting, distinctly papillose, extremely papillose in upper half. Short cilia 2–3 or

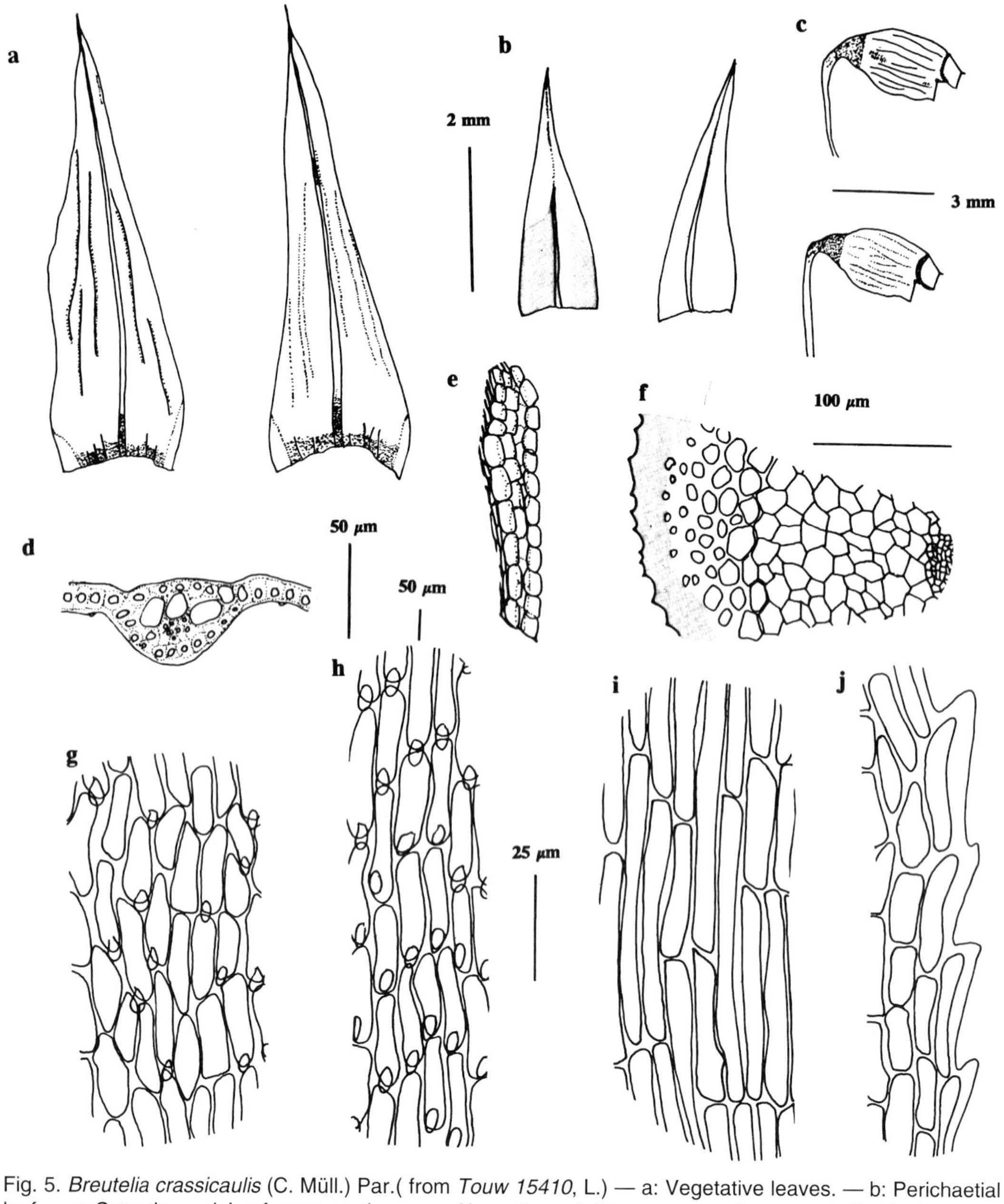

Fig. 5. *Breutelia crassicaulis* (C. Müll.) Par.(from *Touw 15410*, L.) — a: Vegetative leaves. — b: Perichaetial leaf. — c: Capsule. — d: Leaf cross-section. — e: Alar cells. — f: Stem cross-section. — g: Apical leaf cells. — h: Median leaf cells. — i: Basal leaf cells. — j: Apical leaf margin. — Use the 2 mm scale for a–b, the 3 mm scale for c, 50 µm scale on left for d and on right for e, the 100 µm scale for f, and 25 µm scale for g–j.

lacking. Spores ca. 20 µm in diameter, coarsely verrucate.

Illustrations: Fleischer 1904: 631 (fig. 120); Bartram 1933: 143 (fig. 105), 1939: pl. 12 (fig. 207); Noguchi 1989: 575 (fig. 255).

Breutelia arundinifolia is mostly such a robust plant that it is easily identified, but small plants can be confused with *B. roemeri*. The fairly large leaves are, however, a good diagnostic character even in these smaller plants. There is quite much variation in the cell size and form. There are specimens with fairly papillose, rectangular, short cells (as in the type material). Specimens with longer, more linear cells with more distinct papillae resembling *B. roemeri* were illustrated by Bartram (1933). I have seen specimens with apical cells 45 µm long and linear and other specimens with rectangular cells less than 25 µm long. However, I hesitate to give this variation formal taxonomic recognition because of the lack of congruence with other characters.

On the Huon Peninsula *Breutelia arundinifolia* was collected in montane rainforest, open grassland with scattered treeferns, open *Nothofagus* forest, elfin forest, and subalpine forest at the altitudes of 2 750 to 3 250 m. The plants grew mostly epigaeic, and the substrates were humus (3 specimens), peat (2), soil (2), and cliff (1). The record of *B. arundinifolia* from Tasmania (Zanten 1964, Noguchi 1989) is doubtful because the area is far outside of the main distribution area of *B. arundinifolia,* and the species has not been recorded between these two areas. I have not seen any specimens from Tasmania and Zanten (1964) and Noguchi (1989) did not cite any specimens. Also Miller *et al.*(1978), and Streimann and Curnow (1989) regarded the record for Tasmania as doubtful.

Range on the Huon Peninsula: *1a.* 34650. *1b.* 34673. *2e.* 33897. *6a.* 31637, 31723. *6c.* 32029.

Range in Western Melanesia: West Irian. *7.* Papua New Guinea. *10.* Mt. Burgers, grassland, in mud along streamlet, 3 600 m, *Veldkamp & Vinas 7519A* (L). *12.* Mt. Kaindi, disturbed *Nothofagus* forest, on soil on road bank, *Shea 6505* (H). *13.* Wabag area, Ambum-Murimuni divide, open track in montane rain forest, terrestrial, *Robbins 3031* (L). *15.* Mt. Wilhelm, morainic ridge, along small gully, *Balgooy 746A* (L). *17.* Grasslands on gently sloping valley, 2 660 m, *Streimann 32361* (L).

Total range (Bartram 1933, Miller *et al.* 1978, Noguchi 1989): As 2: Ja Taiwan; As 4: Ind Phi PNG; Oc: Haw; Austr 1: Tasmania?.

4. *Breutelia roemeri* Fleisch. (Figs. 4, 7, 9)

Hedwigia 50: 281. 1911. — Type: Indonesia. West Irian, Jayawijaya, Mt. Hellwiggebirage ("Mt. Hellwigis"), Agalhodsmonsberg, 2 500–3 000 m, *Römer 1262* (FH!).

Plants fairly robust, 3–15 cm high, brown, yellow or green, in loose tufts, distinctly tomentose below, ascending, simple or remotely dichotomously branched. Leaves crowded, widely spreading from a subclasping base, often appressed below, mostly with reddish basal part, 3.4–4.2 mm long and 1.1–1.4 mm at widest point, linear-lanceolate, four to five deep plicae at base which can reach up the leaf. Margins recurved near base, entire below, denticulate more than halfway down, apical teeth sharp. Costa narrow, strongly papillose, 3–4 cells broad, red, with clearly excurrent, ca. 330 µm long denticulate apex. Cells linear, the basal parts distinctly papillose, papillae at the base of the cells. Apical cells 22.5–37.5 µm, median cells 25–50 µm, and basal cells 47–62.5 µm long and 2.5–5 µm wide. Alar cells hyaline, in 3–4 rows and reaching up to 6–10 cells, quadrate to rectangular, 30–50 µm long and 20–25 µm wide.

Possibly monoicous. Perigonial leaves ovate to triangular, ca. 4.4 mm long and 2.6 mm wide, basal part red with hyaline cells by margins, costa red basally, green above. Perichaetical leaves lanceolate, ca. 3.4 mm long and 1.3 mm wide basally. Sporophytes rare. Seta 2.5–4 mm long, red to nearly black, capsule ovoid, horizontal, ca. 4 mm long and 2 mm wide. Exostome teeth lanceolete, ca. 0.5 mm long and 0.14 mm wide basally, endostome of ca. 160 µm high basally membrane, from which 250–390 µm long segments arise, segments strongly papillose in upper half, splitting, two short cilia between. Spores verrucate, ca. 23 µm in diameter.

Illustrations: Fleischer 1912: pl. 121 (fig. A).

All *Breutelia* species are mentioned to be dioicous but in one specimen of *B. roemeri* there were both sporophytes and antheridia, on different branches.

Breutelia roemeri can easily be confused with *B. arundinifolia* when the characters are not expressed in their typical form. Mostly *B. roemeri* is not as robust as *B. arundinifolia* and the leaves are smaller, and cells longer and much more strongly papillose than those of *B. arundinifolia.* Also the leaf cells of *B. arundinifolia* are more rectangular especially in the apical part.

Breutelia roemeri is a common plant in New Guinea, and has been collected from open grassland to mossy forest. It has often been found in bogs and also in disturbed sites by road cuttings. It is mostly terrestrial but has rarely been collected as epiphyte. On the Huon Peninsula *B. roemeri* occurred in open grasslands with scattered treeferns (with limestone cliffs), in montane rainforest and moss forest, and in bogs and swamps at the altitudes of 1 500 to 3 400 m. The substrates were boulder (2 specimens), cliff (1), humus (1), peat (4), soil (7) and sinkhole of limestone (1).

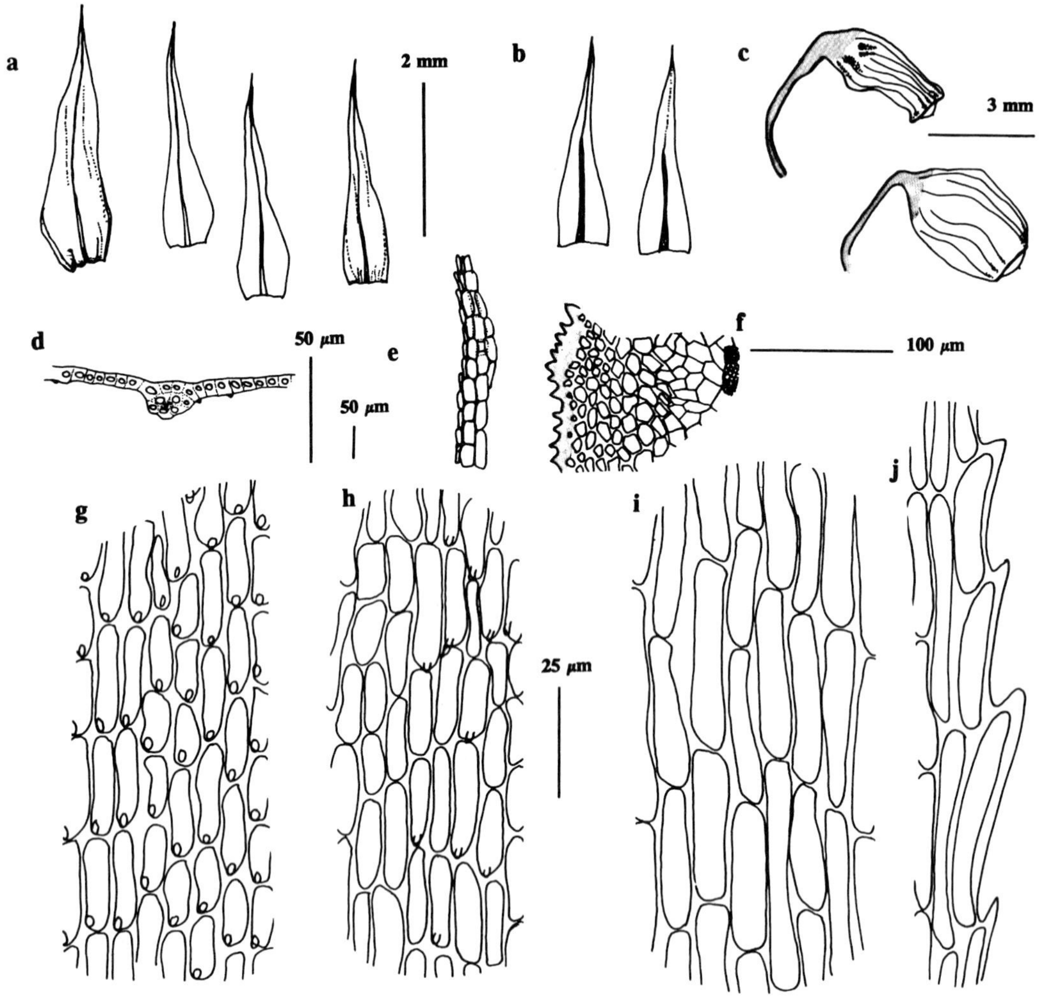

Fig. 6. *Breutelia papuensis* Virtanen (from *Hoogland & Schodde 7026,* holotype, L). — a: Stem leaves. — b: Perichaetial leaf. — c: Capsule (dry above, wet below). — d: Leaf cross-section. — e: Alar cells. — f: Stem cross-section. — g: Apical leaf cells. — h: Median leaf cells. — i: Basal leaf cells. — j: Apical leaf margin. — Use 2 mm scale for a–b, 3 mm scale for c, 50 µm scale on left for d and on right for e, 100 µm scale for f, and 25 µm scale for g–j.

Range on Huon Peninsula: *2d.* 33632. *3a.* 29374. *3h.* 60214. *6a.* 31660. *6e.* 32221. *6i.* 32528. *6j.* 63303. *6k.* 32696. *6u.* 62742B. *6w.* 62815. *7a.* 65310. *8f.* 31311. 31312. 31325. 31330. *8i.* 31517. *10e.* 28804. *10n.* 59757.

Range in Western Melanesia: West Irian. *2.* Mt. Nettoti, in *Sphagnum* cushions in *Nothofagus*-forest, 1 900 m, *Royen 3904* (L). *6. 7.* Papua New Guinea. *12.* Mt. Kaindi, advanced secondary forest, on moist shaded road cutting, 1 360 m, *Streimann 17702* (H). *13. 14.* Milep area, Ming District, epiphyte on mossy forest, *Vinas 7648* (H). *15. 16.* Mt. Otto, terrestrial in mossy forest, 3 500 m, *Veldkamp & Obedi 8709A* (L). *17.* Yebi, Mount Hagen-Mendi road, on grasslands, on moist road cutting, 1 800 m, *Streimann 22126* (H). *20.* Mt. Scratchley, upper montane forest, in shade, very wet area, on forest floor, with *Libertia* and a tussock grass, 3 300 m, *Coode 3814* (L).

Total range: Endemic to New Guinea.

5. *Breutelia crassicaulis* (C. Müll.) Par. (Figs. 5, 8)

Ind. Bryol. Suppl. 53. 1900. — *Bartramia crassicaulis* C. Müll., Flora 82: 447. 1896. — Type: USA. Hawaii. West Maui, 5 700 m, *Hillebrandt* (FI, not seen).

Plants robust, 3–10 cm high, golden brown, in loose tufts, fairly sparse branched. Leaves crowded, erect to erecto-patent, reddish basally, 5.7–7.2 mm long, linear-lanceolate, deeply plicate. Margin entire below, denticulate more than halfway down, teeth

Fig. 7. Habits. — a: *Breutelia aristifolia* Zanten (sporophyte from *Veldkamp & Stevens 5845*, L; the other from *Koponen 32850*). — b: *Breutelia longicapsularis* Dix. (from *Koponen 32559*.) — c: *Breutelia roemeri* Fleisch. (from *Koponen 31325* sporophyte; the other from *Koponen 31660*). — d: *Breutelia arundinifolia* (Duby) Fleisch. (young sporophyte with operculum from *Veldkamp & Vinas 7519A*, L; old sporophyte from *Norris 65209*; sterile from *Koponen 34673*). 68% of natural size. Photo Tuomo Niemelä.

sharp apically but indistinct because of the apically incurved margins. Costa 3–4 cells broad, red basally and yellow above, percurrent to excurrent with fairly short ca. 200 µm long apex. Cells mostly linear, in apex occassionally rectangular, papillose, papillae basal. Apical cells 14–25 µm, median cells 20–37 µm, and basal cells 32–75 µm long, and 2.5–5 µm wide. Alar cells hyaline, in distinct group, 4 rows wide and 6–10 cell-rows long, quadrate to rectangular, 40–60 µm long and 25–37.5 µm wide.

Dioicous. Perichaetial leaves lanceolate, ca. 4.4 mm long and 0.8 mm wide basally, distinct change in colour of cells in the middle of the leaf; red in basally and pale green by apex. Seta ca. 3.5 cm long, red, paler apically, capsule ovoid, 5–6 mm long, 2–3 mm wide, green. Operculum rostellate to mamillate. Exostome teeth ca. 0.58 mm long and 0.17 mm wide basally. Endostome segments were broken in the studied material. Spores smooth, ca. 15 µm in diameter.

The type specimens were not available but I studied two specimens (*Baldwin 582, Baldwin 24*, H-BR) collected during the same expedition as the type specimen and identified by Brotherus, and the other specimen (*Baldwin 582*) possibly with leaves from the original type.

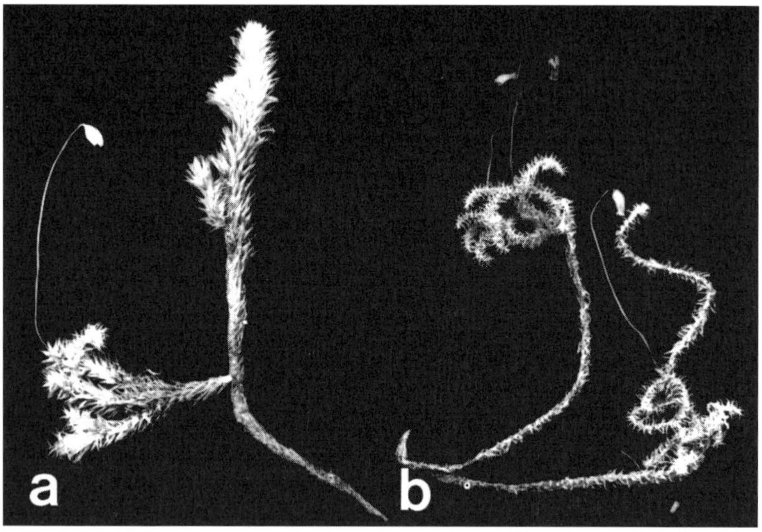

Fig. 8. Habits. — a: *Breutelia crassicaulis* (C. Müll.) Par. (from *Touw 15410,* L). — b: *Breutelia papuensis* Virtanen (from *Hoogland & Schodde 7026*, L). 68% of natural size. Photo Tuomo Niemelä.

Breutelia crassicaulis is the only species of the genus in New Guinea with leaves erect to erecto-patent and therefore it is easily recognized. I have found very few data of *B. crassicaulis* other than the original description. Brotherus (1927) mentioned it from Hawaii where the type material was collected. Bartram's (1939) statement was that he could not recognize any difference between *B. arundinifolia* and *B. crassicaulis*, but there is clear difference in leaf posture: *B. arundinifolia* has leaves spreading and *B. crassicaulis* erect. The subclasping base of leaves, character which is present in the other *Breutelia* species of New Guinea, is lacking in *B. crassicaulis* and because of this character Brotherus (1904) classified *B. crassicaulis* to the section *Lycopodiobryum* (C. Müll.) Broth.

Surprisingly few specimens of *Breutelia crassicaulis* were available for the study although it is quite a robust species. All the specimens I have seen are from swampy (or boggy) sites, at the altitudes of 2 000 m to 2 900 m. It might really be a rare species or the sites are not easily accessible.

Range on the Huon Peninsula: None.

Range in Western Melanesia: West Irian. *7*. Star Mts., Tel Basin Mountain grassland area, small valley filled up with golden coloured *Ectropothecium* bog, terrestrial, 2 950 m, *Touw 15410* (L). Papua New Guinea. *18*. Mt. Karoma, radio station, 3 550 m, recently disturbed area with little grassland in otherwise dense mossy forest on sandy limestone and peat, W slope, on steep limestone rockface, *Veldkamp & Wiakabu 7743B* (L).

Total range: As 4: PNG; Oc: Haw.

6. *Breutelia papuensis* Virtanen, *sp. nov.* (Figs. 6, 8)

Species Breuteliae roemeri *affinis, sed statura minoriore et cellulis apicalibus foliorum minus quam 10 µm longis facillime dignoscenda.*

Holotype: Papua New Guinea. Western Highlands, Wabag subdistrict, northern slopes of Sugarloaf (along Wapu river), in treefern grassland, 3 100 m (c. 9 500 ft), terrestrial, 1960 *Hoogland & Schodde 7026* (L).

Plants fairly robust, 8 cm high, light green, in loose tufts, very distinctly tomentose except the upper 2 to 3 cm, irregularly and sparsely branched, subperichaetial innovations numerous. Leaves lanceolate, crowded, widely spreading from a subclasping base, deep plicates reaching up from the base, basally red. Stem leaves ca. 3.6 mm long, branch leaves < 2.6 mm. Margins entire below, denticulate more than halfway down, apical teeth fairly sharp. Costa narrow, 3–4 cells broad, red, with clearly excurrent, ca. 200 µm long, strongly papillose apex. Cells linear basally, gradually more rectangular towards apex, with distinct basal papillae. Apical cells 12.5–27.5 µm, median cells 17.5–35 µm, and basal cells 37.5–60 µm long, and 3.5–5 µm wide. Alar cells hyaline, in distinct group, cells in 3–4 rows and reaching up to 6 to 8 (– 10) cells, rectangular, 30–50 µm long and ca. 25 µm wide.

Dioicous. Perichaetical leaves linear-lanceolate, ca. 2.3 mm long and 0.4 mm wide basally, costa red and strong basally, pale green and weak by the apex. Seta smooth, 2.5–3 cm long, red, paler apically, cap-

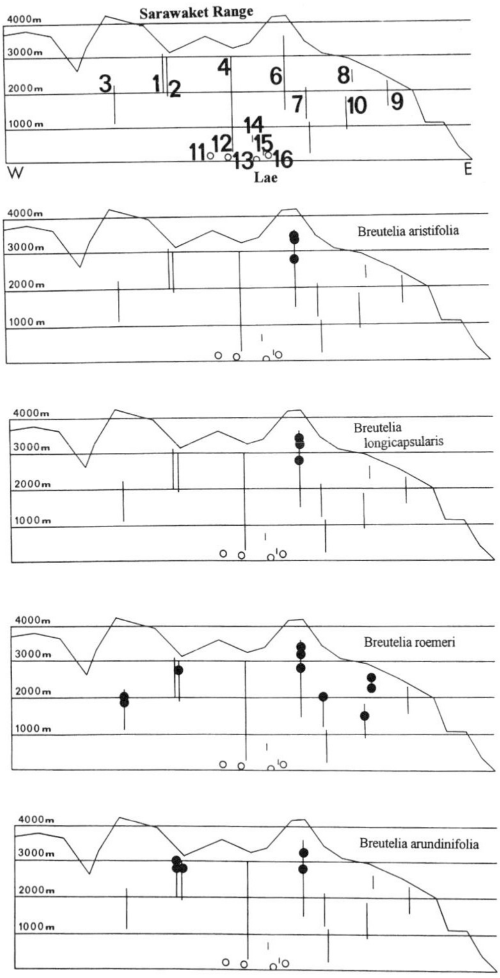

Fig. 9. The altitudinal range of *Breutelia aristifolia* Zanten, *B. longicapsularis* Dix., *B. roemeri* Fleisch., and *B. arundinifolia* (Duby) Fleisch. on the Huon Peninsula. Compare with fig. 4 in Norris & Koponen (1983). Nos. 1–16 indicate the collecting localities and the altitudinal range of the Koponen and Norris expedition.

sule horizontal to pendulous, red or green, ovoid to globose when wet, cylindrical and strikingly furrowed when dry, ca. 4 mm long and 2–3 mm wide when moist, operculum mamillate. Exostome teeth ca. 0.58 mm long and 0.13 mm wide basally, endostome with 200–300 µm high basal membrane, endostome segments ca. 50 µm shorter than exostome teeth, segments strongly papillose in upper half, splitting, no cilia. Spores verrucate (immature ones smooth), ca. 23 µm in diameter.

In SEM pictures the spores of *Breutelia roemeri*, *B. arundinifolia* and *B. papuensis* seemed to resemble each other, but observed with light microscope many spores of *B. papuensis* appeared colorless and smooth whereas all the spores of the two other species were brown and verrucate.

Breutelia papuensis resembles *B. roemeri*, and the habit resembles small plants of *B. roemeri* with abundant branches. However, the cells of *B. papuensis* are shorter than those of *B. roemeri*, and especially the apical cells of *B. papuensis* are clearly rectangular and short compared to the more linear and longer cells of *B. roemeri*.

Range on the Huon Peninsula: None.

Range in Western Melanesia (paratypes): West Irian. *7.* Star Mts., Tel Basin, limestone hill, terrestrial and on rocks, near waterfall, exposed, 3 050 m, *Touw 16002* (L). Tel Basin, on moist perpendicular limestone cliffs in treefern-grassland, 3 000 m, *Touw 15931* (L). Papua New Guinea. *12.* Mt. Bangeta, on ground in montane forest, 24.10.1970 *Argent* (H). *13.* Holotype. *16.* Mt. Otto, summit area, terrestrial in shaded gully in grassland, 3 500 m, *Veldkamp & Obedi 8665B* (L).

Total range: Endemic to New Guinea.

Acknowledgements. I thank my supervisors Prof. Jaakko Hyvönen and Prof. Timo Koponen for patient guidance, and Dr. Johannes Enroth for help with SEM and for the Latin. Prof. Tuomo Niemelä is thanked for taking the photos. Curators of BM, FH, G, H and L are thanked for loaning the specimens.

REFERENCES

Akiyama, H. & Nishimura, N. 1993: The Climacium-type branch development in the Bartramiaceae and its taxonomic significance. — Bryologist 96: 185–191.

Bartram, E. B. 1933: Manual of Hawaiian mosses. — Bishop. Mus. Bull. 101: 1–275.

Bartram, E. B. 1939: Mosses of the Philippines. — Philippine J. Sc. 68(1–4): 1–437.

Brotherus, V. F. 1904: Bartramiaceae. — In: Engler, A. & Prantl, K. (eds.), Die natürlichen Pflanzenfamilien 1(3): 631–660. W. Engelmann, Leipzig.

Brotherus, V. F. 1927: Hawaiian mosses. — Bishop Mus. Bull. 40: 1–37.

Crosby, M. R., Magill, R. E. & Bauer, C. R. 1992: Index of Mosses 1963–1989. — Missouri Bot. Garden, St. Louis. 646 pp.

Crosby, M. R. & Magill, R. E. 1994: Index of Mosses 1990–1992. — Missouri Bot. Garden, St. Louis. 87 pp.

Dixon, H. N. 1922: The mosses of the Wollaston Expedition to Dutsch New Guinea, 1912–1913; with some additional mosses from British New Guinea. — J. Linn. Soc. Bot. 45: 477–510.

Fleischer, M. 1904: Die Musci der Flora von Buitenzorg (zugleich Laubmoosflora von Java) 2: I–XVIII, 381–643. E. Brill, Leiden.

Fleischer, M. 1912: Laubmoose. — In: Résultats de l'Expédition scientifique néerlandaise á la Nouvelle-Guinée. — Nova Guinea 8: 735–753.

Griffin III, D. 1984: Breutelia in Brazil with notes on the occurrence of the genus in the New World. — J. Hattori Bot. Lab. 57: 83–95.

Griffin III, D. & Buck, W. R. 1989: Taxonomic and phylogenetic studies on the Bartramiaceae. — Bryologist 92: 368–380.

Hirohama, T. 1977: Spore morphology of bryophytes observed by scanning electron microscope, II. Bartramiaceae. — Bull. Nat. Sc. Mus. Series B (Botany) 3: 37–44.

Hirohama, T. & Iwatsuki, Z. 1980: Surface ornamentatiom of rhizoids of the species of Bartramiaceae (Musci). — J. Hattori Bot. Lab. 48: 259–275.

Kabiersch, W. 1937: Studien über die ostasiatischen Arten einiger Laubmoosfamilien II (Rhizogoniaceae, Bartramiaceae, Aulacomniaceae, Meesiaceae). — Hedwigia 77(2–3): 71–136.

Kawai, I. 1982: Systematic studies on the conducting tissue of the gametophyte in Musci. (12) Anatomical characteristics of stems in some species of Bartramiaceae. — Sci. Rep. Kanazawa Univ. 26(2): 31–50.

Koponen, T. & Norris, D. H. 1983: Bryophyte flora of the Huon Peninsula, Papua New Guinea. I. Study area and its bryological exploration. — Ann. Bot. Fennici 20: 15–29.

Koponen, T., Piippo, S. & Norris, D. H. 1991: Bryophyte flora of the Huon Peninsula, Papua New Guinea. I–XLVI. Generic and family indexes to parts I–XLVI. — Acta Bot. Fennica 143: 107–111.

Manuel, M. G. 1981: A synopsis of Bartramiaceae Schwaegr. (Bryophytina) in Malaya. — J. Hattori Bot. Lab. 50: 249–252.

Matteri, C. M. 1973: El género Breutelia (Bartramiaceae, Musci) en la región Andino-Patagónica. — Rev. Mus. Argentino Cs. Nat. Bernardino Rivadavia, Bot. 4: 321–360.

Miller, H. A., Whittier, H. O. & Whittier, B. A. 1978: Prodromus florae muscorum Polynesiae; with a key to genera. — Bryophyt. Biblioth. 16: 1–334.

Norris, D. H. & Koponen, T. 1985: Bryophyte flora of Huon Peninsula, Papua New Guinea. VII. Trachypodaceae, Thuidiaceae and Meteoriaceae (Musci). — Acta Bot. Fennica 131: 1–51.

Noguchi, A. (supplemented by Z. Iwatsuki) 1989: Illustrated moss flora of Japan. Part 3: 493–742. — Hattori Bot. Lab., Nichinan.

O'Shea, B. 1995: Checklist of the mosses of sub-Saharan Africa. — Trop. Bryol. 10: 91–198.

Sloover, J. L. De 1975: Note de bryologie Africaine. IV. — Breutelia. — Bull. Jard. Bot. Natl. Belgique 45: 237–271.

Streimann, H. & Curnow, J. 1989: Catalogue of mosses of Australia and its external territories. — Australian Flora and Fauna Series 10: I–VIII, 1–479. Australian Governm. Publ. Serv., Canberra.

Wade, L. K. & McVean, D. N. 1969: Mt Wilhelm Studies I. The alpine and subalpine vegetation. — Research School of Pacific Studies. Dept. of Biogeogr. & Geomorph. Publ. BG/1, Canberra. 225 pp.

Wijk, R. van der, Margadant, W. D. & Florschütz, P. A. 1959. Index Muscorum. Vol. I (A–C). — Regnum Vegetabile 17: 1–548.

Zanten, B. O. van 1964: Mosses of the Star Mountains expedition. — Nova Guinea 16: 263–368.